SpringerBriefs in Electrical and Computer Engineering

Series Editor:
Woon-Seng Gan, School of Electrical and Electronic Engineering
Nanyang Technological University, Singapore, Singapore
C.-C. Jay Kuo, University of Southern California, Los Angeles, CA, USA
Thomas Fang Zheng, Research Institute of Information Technology
Tsinghua University, Beijing, China
Mauro Barni, Department of Information Engineering and Mathematics
University of Siena, Siena, Italy

SpringerBriefs present concise summaries of cutting-edge research and practical applications across a wide spectrum of fields. Featuring compact volumes of 50 to 125 pages, the series covers a range of content from professional to academic. Typical topics might include: timely report of state-of-the art analytical techniques, a bridge between new research results, as published in journal articles, and a contextual literature review, a snapshot of a hot or emerging topic, an in-depth case study or clinical example and a presentation of core concepts that students must understand in order to make independent contributions.

More information about this series at http://www.springer.com/series/10059

Gholamreza Vahedi Sarrigani • Iraj Sadegh Amiri

Willemite-Based Glass Ceramic Doped by Different Percentage of Erbium Oxide and Sintered in Temperature of 500-1100C

Physical and Optical Properties

 Springer

Gholamreza Vahedi Sarrigani
School of Chemical and Biomolecular
Engineering
The University of Sydney
Darlington, NSW, Australia

Materials Synthesis and Characterization
Laboratory
Institute of Advanced Technology
Universiti Putra Malaysia
Serdang, Selangor, Malaysia

Iraj Sadegh Amiri
Ton Duc Thang University
Ho Chi Minh City, Vietnam

ISSN 2191-8112 ISSN 2191-8120 (electronic)
SpringerBriefs in Electrical and Computer Engineering
ISBN 978-3-030-10643-0 ISBN 978-3-030-10644-7 (eBook)
https://doi.org/10.1007/978-3-030-10644-7

Library of Congress Control Number: 2019932164

© The Author(s), under exclusive licence to Springer Nature Switzerland AG 2019
This work is subject to copyright. All rights are reserved by the Publisher, whether the whole or part of
the material is concerned, specifically the rights of translation, reprinting, reuse of illustrations, recitation,
broadcasting, reproduction on microfilms or in any other physical way, and transmission or information
storage and retrieval, electronic adaptation, computer software, or by similar or dissimilar methodology
now known or hereafter developed.
The use of general descriptive names, registered names, trademarks, service marks, etc. in this publication
does not imply, even in the absence of a specific statement, that such names are exempt from the relevant
protective laws and regulations and therefore free for general use.
The publisher, the authors, and the editors are safe to assume that the advice and information in this book
are believed to be true and accurate at the date of publication. Neither the publisher nor the authors or the
editors give a warranty, express or implied, with respect to the material contained herein or for any errors
or omissions that may have been made. The publisher remains neutral with regard to jurisdictional claims
in published maps and institutional affiliations.

This Springer imprint is published by the registered company Springer Nature Switzerland AG
The registered company address is: Gewerbestrasse 11, 6330 Cham, Switzerland

Preface

Zinc silicate glass is an attractive host matrix for rare-earth ions because of its fine properties, primarily optical and mechanical properties, such as good chemical stability, high UV transparency, high surface damage threshold, large tensile fracture strength, and good durability. Up to now, most research has been carried out on soda–lime–silicate (SLS) glass doped with different ingredients and rare-earths, but a few researches have been carried out on willemite-based glass-ceramics prepared using waste material and doped with erbium oxide (Er_2O_3). However, using waste materials such as SLS glass as a main source for producing silicate will be economical, cheap, and helpful for reducing the aggregation of waste materials from the landfill.

The main objective of this study is to determine the effect of erbium oxide (Er_2O_3) addition on physical and optical properties of willemite-based glass-ceramics sintered at different temperatures. The samples were produced via melt-quenching technique followed by powdering, pressing, and sintering. In the first stage, the SLS glasses were crushed, grounded, and sieved to gain the expected particle size. The prepared powder was mixed with zinc oxide (ZnO), followed by melting at the temperature of 1400 °C and quenching in water to obtain fritz glass. The prepared fritz glass was crushed using mortar and pestle to the size of 63 μm. After that, the prepared powder was heat-treated at the temperature of 1000 °C to produce willemite. The willemite-based glass-ceramics was doped with trivalent erbium (Er^{3+}) in the ($[(ZnO)_{0.5}(SLS)_{0.5}]_{1-x}[Er_2O_3]_x$) composition where $x = 1$–5 wt.%. At the end, the powder was pressed, and different pallets were prepared and finally sintered at different temperatures ranged from 500 to 1100 °C. The crystal (phase) changes with different contents of Er_2O_3 and different sintering temperatures were investigated using X-ray diffraction (XRD); the binding structure was explored by Fourier-transform infrared spectroscopy (FTIR); the microstructure, morphology, and chemical composition were being studied using Field Emission Scanning Electron Microscopy with Energy Dispersive X-Ray Spectroscopy (FESEM-EDX); and the optical properties were analyzed by UV–VIS spectroscopy.

The XRD results show that well-crystallized willemite (Zn_2SiO_4) with the contribution of dopant (Er^{3+}) in the lattice can be achieved at the temperature of 900 °C. The

XRD results also show that rhombohedral crystalline willemite was formed by mixing ZnO and SLS glass and optimum heat treatment of 1000 °C to produce willemite-based glass-ceramics, the solid-state reaction between well-crystallized willemite and Er^{3+} was obtained at 900 °C sintering temperature, and Er^{3+} can be completely dissolved in the lattice at this temperature. FTIR results confirmed the appearance of the vibrations of SiO_4 and ZnO_4 groups which clearly suggests the formation of the Zn_2SiO_4 phase; the compositional evaluation of the FTIR properties of the $[(ZnO)_{0.5}(SLS)_{0.5}]_{1-x}[Er_2O_3]_x$ system indicates that the presence of erbium ions affects the surrounding of the Si-O and trivalent erbium occupies their position; these agree with the XRD data at the peak positioned at 20.29°. The most significant modification produced by the addition of erbium and the increase of the heat treatment temperature of the studied samples shows a drop in the intensity of FTIR band located at 513 cm^{-1}, which indicates that the addition of erbium oxide and increase in the sintering temperature decline the presence of SiO_4 group. The microstructure analysis of the samples using FESEM shows that the average grain size of samples tends to increase from 325.29 to 625.2 nm as the sintering temperature increases. Finally, the UV–VIS spectra of all doped glass-ceramics depict absorption band due to host matrix network and the presence of Er_2O_3. The results show that the intensity of the bands tends to grow by increasing the Er_2O_3 content in the range of 1–5 wt.% and the sintering temperature in the 500–900 °C range, followed by a drop at the temperatures of 1000 and 1100 °C. By adding the Er_2O_3 content to the host network and increasing the sintering temperature from 500 to 900 °C , the intensity of UV–VIS bands situated between 400 and 1800 nm increased due to the absorption of Er^{3+}ions and the host crystal structure. The intensity of the UV bands was observed to have dropped when the sintering temperature was increased to 1000 and 1100 °C, which indicates that by going to the temperature of 1000 and 1100 °C, the Er_2O_3 particles tend to produce cluster that causes the decrease in the UV absorption bands. For the sample with x = 5 wt.% Er_2O_3, two strong absorption bands situated at about 1535 and 523 nm were observed. These bands were attributed to the optical transition from $^4I_{15/2}$ to $^4I_{13/2}$ and $^4S_{3/2}$ state, respectively.

Darlington, NSW, Australia   Gholamreza Vahedi Sarrigani
Ho Chi Minh City, Vietnam   Iraj Sadegh Amiri

Contents

Chapter 1
Introduction to Glass and Glass-Ceramic Background

1.1 Glass and Glass-Ceramic Materials

Glass is a product of inorganic fusion obtained by cooling down molten inorganic materials to a rigid condition. Glasses can be synthesized in various shapes by the melting and quenching method. Today, the uses of glasses are far ranging, with applications in the architectural, electrical and electronic device, telecommunications, and aerospace industries. Generally glass ceramics are known to have an amorphous phases and one or more crystalline phases. Glass ceramics are produced through a controlled process of crystallization of the base glass, in contrast to spontaneous crystallization, which is not acceptable to glass manufacturers. Glass-ceramic materials exhibit properties of both glasses and ceramics, including the fabrication advantage of glass and special properties of ceramics. Glass ceramics usually have crystallinity between 30% [m/m] and 90% [m/m] [1]. Glass-ceramic materials have properties of high strength, translucency, pigmentation, opalescence, high chemical durability, high temperature stability, low or negative thermal expansion, fluorescence, machinability, ferromagnetism, resorbability, biocompatibility, bioactivity, ion conductivity, superconductivity, isolation capabilities, a low dielectric constant and low dielectric loss, high resistivity, and a high breakdown voltage [2, 3]. These properties can be optimized by controlling the composition of the base glass and applying a controlled heat treatment/crystallization to the base glass.

Most production of glass ceramics is done in two steps: first, a glass is produced through a glass-manufacturing process such as melting and quenching. Then the glass is reheated again at a specific temperature. During this heat treatment, the glass undergoes partial crystallization. The properties of glass ceramics are determined by the precipitation of crystallized phases from the glasses, as well as their microstructure. Generally, control of the crystallinity and the type of crystal structure in the final glass ceramics depend on the parent glass composition, thermal

© The Author(s), under exclusive licence to Springer Nature Switzerland AG 2019
G. V. Sarrigani, I. S. Amiri, *Willemite-Based Glass Ceramic Doped by Different Percentage of Erbium Oxide and Sintered in Temperature of 500-1100C*,
SpringerBriefs in Electrical and Computer Engineering,
https://doi.org/10.1007/978-3-030-10644-7_1

treatment, and addition of a nucleating agent [4]. Mostly, nucleating agents are added to the base composition of the glass ceramics to control and facilitate the crystallization process. Nucleation is the initiation of a phase change in a small region, such as the formation of a solid crystal from a liquid solution. It is a consequence of rapid local fluctuations on a molecular scale in a homogeneous phase that is in a state of metastable equilibrium.

1.2   Background on Sintering and Sintering Temperature

Sintering is a technique used to produce substances from powders, which is based on diffusion of atoms. Diffusion takes place at a temperature above absolute zero in any material; however, at an elevated temperature the diffusion is much quicker. Sintering is the process of heating the powdered material to a temperature below the melting point. During sintering processes, the atomic composition of the powdery particles results in diffusion across the boundaries of the particles, producing a solid piece by fusing the particles together. Sintering is known as a shaping process for materials, such as glass and glass ceramics, with extremely high melting points. Besides that, sintering is part of the process used for manufacturing pottery and ceramic objects and substances such as glass, alumina, zirconia, silica, magnesia, lime, beryllium oxide, and ferric oxide. The benefits of the sintering stage are higher levels of purity and uniformity in the starting materials and preservation of purity, since it has simpler and fewer subsequent fabrication steps [5, 6]. Stability of the details in repeatable operations is achieved by control of the grain size at the input stages; there is no binding contact between segregated powder particles as is often observed in processes of melting. There are different benefits of sintering, such as no deformation being needed to produce directional elongation of grains, ability to achieve materials with controllable and uniform porosity, ability to obtain objects with a near-net shape, ability to achieve materials that are difficult to fabricate by other techniques, and ability to produce highly durable material.

A higher sintering temperature will cause glass-ceramic particles to be compacted. It tends to decrease the porosity and increase the density, as well as the grain size, of the glass ceramic. After sintering, the glass ceramic has higher thermal conductivity, a higher elastic modulus, and greater mechanical strength [7]. Sintering also increases the diffusion rate and reduces the dislocation of particles. The effective sintering temperature for the microstructure of a single-phase oxide is in the range of 0.75–0.90 of its melting point [7]. Effective solid-state sintering is also limited to powder with a particle size less than 10 μm, because finer powder has a greater surface area per unit of volume. Consequently, it has a greater driving force to cause densification at lower temperature [7]. The effectiveness of a solid-state sintering process can also be increased by application of higher pressure, slowing the heating rate, and extending the heating period. There are various mass transport mechanisms involved in solid-phase sintering of powder, which include surface diffusion, volume diffusion, grain boundary diffusion, viscous flow, plastic flow, and vapor transport from a solid surface [8].

There are three stages of solid-phase sintering that take place in microstructure evolution. The stages of the solid-phase sintering depend on the sintering temperature and the sintering period, as well as the nature of the material, such as the melting temperature and the particle size, shape, and surface. Assuming that all of the powder particles are spherical, there is wide interparticle spacing present in the powder. The first stage of sintering occurs when the spherical particles come into contact, with a weak cohesive force within them. A small mass of particles is participated to form a neck. Hence, the neck growth of the particles and decreased porosity are significant in microstructures [8]. The intermediate stage of sintering occurs as a large mass of particles is involved in the neck growth. The particles are no longer spherical because they are interconnected. An open pore network with porosity greater than 8% becomes geometrically unstable. The pores undergo shrinkage and become smoother. At this stage, significant densification of the microstructure can be observed [8].

During the final stage of sintering, grain growth and densification are evident. Several grains are joined together, and growth to a larger size occurs. Hence, the average grain size is increased and fewer grains can be observed in a unit area of a micrograph. The pores observed are spherical and closed. They are present on the fractured grain boundaries. The total surface porosity achieved is lower than 8%. The air in the pores will limit the end points of the total porosity and density after sintering [8].

1.3 The Concept of Doping

The doping process was first properly developed by John Robert Woodyard at Sperry Gyroscope Company during World War II [9]. Other research in line with his work was also performed by Teal and Sparks at Bell Labs [10].

Generally, doping is the process of adding impurities to a material. For instance, in the fabrication of semiconductors, impurities are usually introduced into the host lattice to modify its electrical and optical properties. Doping processes are mainly important for the creation of electronic junctions in silicon and for the manufacturing of semiconductor devices [11]. In the technology of glass-ceramic phosphor materials, some impurities such as rare-earth elements are utilized as activators to enhance the luminescence characteristics. Such phosphor materials, prepared from inorganic compounds by doping with suitable activators and impurities, are capable of converting one or more forms of energy into radiation in or close to the visible region of the electromagnetic spectrum. Erbium-doped materials have been widely studied for several years, as trivalent erbium (Er^{3+}) ions illustrate emission at 1500 nm, which coincides with the minimum-loss transmission window of silica-based optical fibers used in telecommunication systems [12].

1.4   Rare-Earth Luminescence in Solid Hosts

The rare earths, or lanthanides, are the series of elements in the sixth row of the periodic table, starting with lanthanum and ending with ytterbium. Rare earths are specially identified by a partially filled 4f shell that is shielded from the outer field by $5s^2$ and $5p^6$ electrons. In this series, the energy levels of the elements are not highly sensitive to the surrounding environment they are in.

Rare earths can be incorporated into crystalline or amorphous hosts in the form of 3+ ions or, occasionally, 2+ ions. The 3+ ions all exhibit strong narrow-band intra-4f luminescence in different hosts and shielding provided by the $5s^2$ and $5p^6$ electrons, meaning that rare earths have radiative transitions in solid hosts similar to those of the free ions and weak electron–phonon coupling. Diagrams related to the energy level of the isolated 3+ ions of each of 13 lanthanides with partially filled 4f orbitals, from cerium ($n = 1$) to ytterbium ($n = 13$), are shown in Figs. 1.1 and 1.2. Though some divalent species (principally samarium and europium) also show luminescence, it is the trivalent ions that are mostly of high interest. The intra-4f transitions are parity forbidden and are partially allowed by crystal–field interactions mixing opposite parity wave functions.

Therefore, luminescence has a long lifetime (in the range of milliseconds), with narrow line widths. An intense narrow-band emission can be achieved by choosing suitable ions across most of the visible region and into the near infrared. The more technologically important radiative transitions are highlighted. Figure 1.3 further

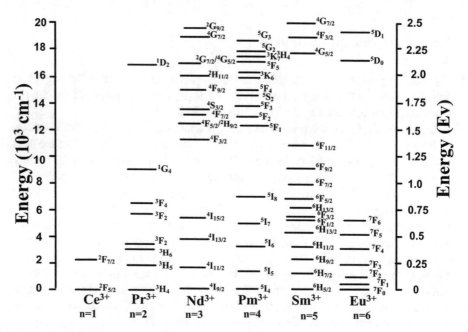

Fig. 1.1 Energy levels of triply charged lanthanide ions ($n = 1–6$)

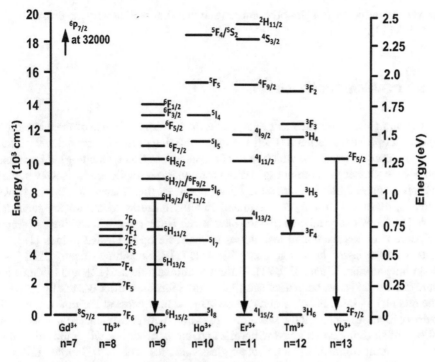

Fig. 1.2 Energy levels of triply charged lanthanide ions ($n = 7$–13)

Fig. 1.3 Influence of spin–orbit and crystal field splitting on the energy levels of the trivalent erbium ion in a silicate host

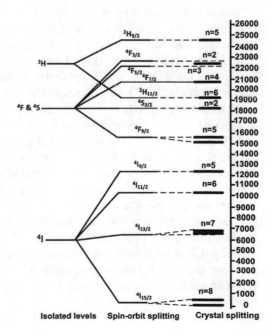

shows the influence of spin–orbit and crystal field interaction with the energy of the Er^{3+} ion [13].

1.5   Problem Statement

Phosphate glasses are excellent host media for Er^{3+} ions because of their attractive spectroscopic characteristics (such as a large emission cross-section and weak inter-action among active ions, which may cause concentration quenching). At present, phosphate glasses are commonly utilized for bulk laser applications. However, they are not very suitable for integrated optics purposes (i.e., planar or channel wave-guides), because of their poor chemical stability and low transition temperatures [14, 15]. Conversely, silicate glasses have much better chemical stability, which is important for ion exchange techniques to fabricate optical wave guides [16]. To date, silicate glasses such as germanosilicate [17, 18], soda lime silicate (SLS) [19], soda-lime aluminosilicate [20, 21], lithium aluminum silicate [22], and lithium sili-cate glasses [23] have been used as suitable hosts for rare earths, especially Er^{3+}. On the other hand, there is only limited knowledge of the effects of erbium on zinc sili-cate (willemite)—especially willemite prepared with waste materials—and the effects of sintering on undoped willemite and rare-earth-doped willemite sintered at different temperatures. In fact, silicate glass ceramics such as willemite (zinc sili-cate), which is a very interesting kind of silicate, are the semiconductors of choice for the overwhelming majority of microelectronics, and full integration of silicate microelectronics with optical emission would allow the realization of low-cost, high-speed communication within circuits, between processors, or across local area networks. Moreover, many countries are facing difficulty nowadays in disposing of solid waste materials from industries and from man-made waste, because of limita-tions in the availability of landfill sites at which to dump this solid waste. In Malaysia, 19,000 tons of waste are produced every day, and a majority of that waste ends up in landfills. Malaysia currently has 230 landfill sites, and 80% of them will reach maximum capacity within the next 2 years. With land for landfill sites being at a premium, there is going to be a big problem for the next generation. Recycling of low-cost waste materials such as SLS glass to fabricate materials applicable to the fields of optics and telecommunications will be beneficial to reduce the large amount of solid waste produced daily. Generally, SLS glasses have been known for their high insulation properties [24], good mechanical properties [25], and accept-able chemical properties. Also, they have been used in radiation-sensitive dosime-ters, especially glasses doped with transition metal ions [26] or rare-earth ions [27]. Besides that, use of waste material in the scientific field is affordable because the majority of waste materials are inexpensive and accessible. The artificial materials that have potential for use in production of glass ceramics are SLS, table salt, and aluminum cans. The main framework of SLS glass is a SiO_4 tetrahedron of silica, which plays an important role as a network-forming oxide. Pure silica has a melting temperature of 1713 °C, which is very high and unaffordable. On the other hand, use of waste SLS is economical because it has soda as a flux, which reduces the eutectic temperature to ~800 °C at the silica-rich end of the phase diagram, and the presence of SLS glass in other oxides is capable of enhancing oxide interaction and crystal formation upon sintering.

1.6 Importance of this Study

Nowadays, glasses and glass ceramics play a major role in telecommunication systems, and they have been intensively studied for applications in conversion fibers, optical amplifiers, solid-state lasers, and three-dimensional displays. Previously, most researches focused on silica oxide as a glass-forming network. Among oxide glasses, phosphate and silicate glasses are the two most important materials, and they have been used extensively for lasers and fiber amplifiers [28]. Compared with silicate glasses, phosphate glasses are more limited in their use because they are hydroscopic in nature and have a lower glass transition temperature. In contrast, silicate glasses exhibit superior chemical resistance and are optically transparent at the excitation and lasing wavelengths [29]. Therefore, they are more compatible with the fabrication process in the development of optical devices.

Various examples of rare earths—such as erbium (Er^{3+}) [17], europium (Eu^{3+}) [30], terbium (Tb^{3+}) [31], and cerium (Ce^{3+}) [32]—are used as dopants in silicate glass ceramics to produce a full-color display. A particularly remarkable use of rare-earth ions is as phosphor activators. In the case of luminescent rare earths, attention has been focused on one species—trivalent erbium (Er^{3+})—with an emission band of around 1.53 mm. The justification for this is evident when one considers the rapid increase in optical telecommunications and some of the material limitations of this technology.

Figure 1.4 shows the loss spectrum of silica fiber. There are two low losses (or windows) in the spectrum of silica fiber: one between 1200 and 1350 nm, and a second one around 1450–1600 nm (known as ultra-low-loss windows). This phenomenon is caused by the combined effects of losses due to Rayleigh scattering and infrared absorption due to the Si–O species. The 1500-nm window is the wavelength region of choice for telecommunications, which fortuitously coincides with the 1535-nm intra-4f $^4I_{13/2}$–$^4I_{15/2}$ transition of the Er^{3+} ion (Fig. 1.5). Therefore, there has been major interest in using erbium-doped materials in telecommunication systems. In the late 1980s, the development of the erbium-doped fiber amplifier (EDFA) [33, 34] exploited the $^4I_{3/2}$–$^4I_{15/2}$ transition and permitted transmission and amplification of signals in the 1530- to 1560-nm region without the need for expensive optical to electrical conversion.

On the other hand, willemite, as a particular kind of silicate glass ceramic, can also be considered a suitable host for erbium [35]. Willemite has been researched over the last 180 years and is still the most widely used, practical, and interesting option for hosting of rare-earth ions. Since its discovery, researchers have focused on the occurrence, crystallography, luminescence, and industrial application of

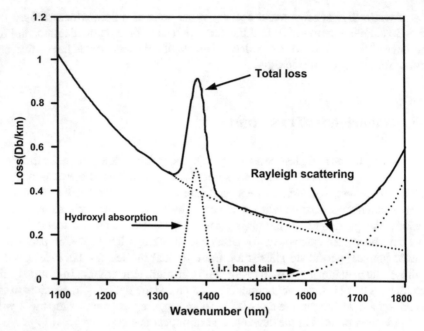

Fig. 1.4 Silicate optical fiber loss spectrum in the near-infrared region

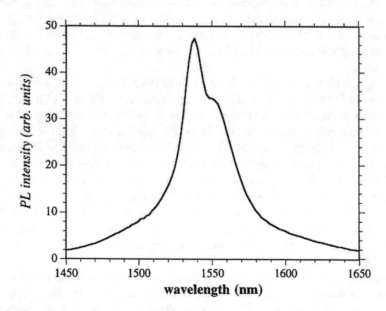

Fig. 1.5 Photoluminescence spectrum of trivalent erbium ($^4I_{13/2}$–$^4I_{15/2}$) in a silica host

willemite. To date, willemite has generally been created using pure materials and different methods such as sol–gel methods, supercritical water methods, vapor methods, and solid-state methods; also, for use in different areas, different kinds of metals and rare earths have been chosen for doping of willemite glass ceramics [36].

At present, there is no published research on production of willemite by use of waste materials and use of Er^{3+} as a dopant. In this work, willemite as a host was produced by using waste material, and Er^{3+} was used as a dopant material.

1.7 Objectives of this Research

This study is based on the following objectives:

1. To analyze the density and linear shrinkage of willemite-based glass ceramic with addition of different percentages of Er_2O_3 and sintered at various temperatures
2. To characterize the crystal phase and microstructure of the produced willemite-based glass ceramic
3. To determine the optimal content of Er_2O_3 and the optimal sintering temperature for willemite-based glass ceramic
4. To evaluate the optical properties of Er_2O_3-doped willemite-based glass ceramic by utilization of ultraviolet–visible (UV-VIS) spectroscopy

1.8 Scope of this Study

To achieve the aims of the study, the scope of the study is as follows:

1. The base of willemite with a stoichiometric equation of $(ZnO)_{0.5}(SLS)_{0.5}$ is produced using a melting and quenching technique followed by sintering, and then the willemite-based glass ceramic is doped with Er_2O_3 from 1 to 5 wt.%.
2. The structure of the samples is investigated using an x-ray diffraction technique to settle the crystalline structure of the glass-ceramic samples.
3. The bonding structure of the samples is detailed by using Fourier transform infrared spectroscopy (FTIR).
4. The density of the obtained samples is investigated using the Archimedes principle with water as the fluid medium.
5. The microstructure, morphology, and chemical composition structure of the samples are studied using field emission scanning electron microscopy (FE-SEM) along with energy-dispersive x-ray analysis (EDAX).
6. The optical properties of the samples are investigated using UV-VIS spectroscopy.

1.9   Organization of this Book

This book contains five chapters. Chapter 1 presents a brief introduction to the research topic and states the objectives of the research. Chapter 2 provides a survey of the literature on glass and glass ceramics, with brief information about willemite preparation. Details of the experimental procedures performed in this work are presented in Chap. 3. Chapter 4 presents the results of x-ray diffraction (XRD) analysis, FTIR spectra, density and linear shrinkage, UV-VIS analysis, and FE-SEM analysis. Chapter 5 contains the conclusions and recommendations for future work and refinement that can be done.

References

1. F.A. Hummel, Thermal expansion properties of some synthetic lithia minerals. J. Am. Ceram. Soc. **34**(8), 235–239 (1951)
2. W. Holand, G. Beall, Crystallization and properties of a Spodumene-Willemite glass ceramic, in *Glass-Ceramic Technology* (Wiley, New York, 2002)
3. P.W. McMillan, The glass phase in glass-ceramics. Glass Technol. **15**(1), 5–15 (1974)
4. A.M. Hu, M. Li, D.L. Dali, K.M. Mao Liang, Crystallization and properties of a Spodumene-Willemite glass ceramic. Thermochim. Acta **437**(1–2), 110–113 (2005)
5. S.L. Kang, *Sintering: Densification, Grain Growth, and Microstructure*, vol 5 (Elsevier, Amsterdam, 2005), pp. 9–18
6. C.B. Carter, M.G. Norton, *Ceramic Materials: Science and Engineering*, vol 9 (Springer, New York, 2007), pp. 427–443
7. J.F. Shackelford, R.H. Doremus, *Ceramic and Glass Materials, Structure, Properties and Processing* (Springer, New York, 2008), pp. 1–209
8. R.M. German, *Sintering Theory and Practice* (Wiley, New York, 1996)
9. J.R. Woodyard, *Electrical Engineering*, vol 21 (University of California, Berkeley, 1985), pp. 8–12
10. M.A.T. Sparks, K. Gordon, Method of making P-N junctions in semiconductor materials, 1950 U.S. Patent 2,631,356, 17 March, 1953
11. S.M. Sze, *Physics of Semiconductor Devices* (Wiley, New York, 1981)
12. W.J. Miniscalco, Erbium-doped glasses for fiber amplifiers at 1500 nm. J. Lightwave Technol. **9**, 234–250 (1991)
13. A.J. Kenyon, Recent developments in rare-earth doped materials for optoelectronics. Prog. Quantum Electron. **26**(4–5), 225–284 (2002)
14. J. Yang, S. Dai, N. Dai, L. Wen, L. Hu, Z. Jiang, Investigation on nonradiative decay of $^4I_{13/2} \rightarrow {}^4I_{15/2}$ transition of Er^{3+}-doped oxide glasses. J. Lumin. **106**, 9–14 (2004)
15. C. Tanaram, C. Teeka, R. Jomtarak, P.P. Yupapin, M.A. Jalil, I.S. Amiri, J. Ali, ASK-to-PSK generation based on nonlinear microring resonators coupled to one MZI arm. Proc. Eng. **8**, 432–435 (2011)
16. P. Capek, M. Mika, J. Oswald, P. Tresnakova, L. Salavcova, O. Kolek, J. Spirkova, Effect of divalent cations on properties of Er^{3+}-doped silicate glasses. Opt. Mater. **27**(2), 331–336 (2004)
17. R. Santos, L. Santos, R. Almeida, Optical and spectroscopic properties of Er-doped niobium germanosilicate glasses and glass ceramics. J. Non-Cryst. Solids **356**(44–49), 2677–2682 (2010)
18. R.A. Smith, *Semiconductors Second Edit* (Cambridge University Press, Cambridge, 1978)

19. E.M.A. Khalil, F.H. ElBatal, Y.M. Hamdy, H.M. Zidan, M.S. Aziz, A.M. Abdelghany, Infrared absorption spectra of transition metals-doped soda lime silica glasses. Phys. B Condens. Matter **405**(5), 1294–1300 (2010)

20. S. Berneschi, M. Bettinelli, M. Brenci, R. Dall'Igna, G. Nunzi Conti, S. Pelli, B. Profilo, S. Sebastiani, A. Speghini, G.C. Righini, Optical and spectroscopic properties of soda-lime alumino silicate glasses doped with Er^{3+} and/or Yb^{3+}. Opt. Mater. **28**(11), 1271–1275 (2006)

21. G.C. Righini, C. Arnaud, S. Berneschi, M. Bettinelli, M. Brenci, A. Chiasera, L. Zampedri, Integrated optical amplifiers and microspherical lasers based on erbium-doped oxide glasses. Opt. Mater. **27**(11), 1711–1717 (2005)

22. A. Ananthanarayanan, G.P. Kothiyal, L. Montagne, B. Revel, MAS-NMR investigations of the crystallization behaviour of lithium aluminum silicate (LAS) glasses containing P_2O_5 and TiO_2 nucleants. J. Solid State Chem. **183**(6), 1416–1422 (2010)

23. J. Du, C. Chen, Structure and lithium ion diffusion in lithium silicate glasses and at their interfaces with lithium lanthanum titanate crystals. J. Non-Cryst. Solids **358**(24), 3531–3538 (2012)

24. Y. Hayashi, M. Kudo, Mechanism for changes in surface composition of float glass and its effects on the mechanical properties. J. Chem. Soc. Jpn. Chem. Ind. Chem. **4**, 217–228 (2001)

25. H. Wang, G. Isgrò, P. Pallav, A. Feilzer, J. Chao, Y. Chao, Influence of test methods on fracture toughness of a dental porcelain and a soda lime glass. J. Am. Ceram. Soc. **88**(10), 2868–2873 (2005)

26. C. Mercier, G. Palavit, L. Montagne, C. Follet-Houttemane, A survey of transition-metal-containing phosphate glasses. C. R. Chim. **5**(11), 693–703 (2002)

27. F.H.A. Elbatal, M.M.I. Khalil, N. Nada, S.A. Desouky, Gamma rays interaction with ternary silicate glasses containing mixed CoO+NiO. Mater. Chem. Phys. **82**(2), 375–387 (2003)

28. D.L. Veasey, D.S. Funk, N.A. Sanford, J.S. Hayden, Arrays of distributed-Bragg-reflector waveguide lasers at 1536 nm in Yb/Er co-doped phosphate glass. Appl. Phys. Lett. **74**, 789–791 (1999)

29. H. Lin, E.Y.B. Pun, X.R. Liu, Erbium-activated aluminum fluoride glasses: optical and spectroscopic properties. J. Non-Cryst. Solids **283**, 27–33 (2001)

30. J. Du, L. Kokou, Europium environment and clustering in europium doped silica and sodium silicate glasses. J. Non-Cryst. Solids **357**(11–13), 2235–2240 (2011)

31. Z. Pan, K. James, Y. Cui, A. Burger, N. Cherepy, S.A. Payne, S.H. Morgan, Terbium-activated lithium–lanthanum–aluminosilicate oxyfluoride scintillating glass and glass-ceramic. Nucl. Instrum. Methods Phys. Res., Sect. A **594**(2), 215–219 (2008)

32. A. Brandily, L. Marie, J. Lumeau, L. Glebova, L. Glebov, Specific absorption spectra of cerium in multicomponent silicate glasses. J. Non-Cryst. Solids **356**(44–49), 2337–2343 (2010)

33. E. Desurvire, R.J. Simpson, P.C. Becker, Preparation and characterizations of Zn_2SiO_4. Opt. Lett. **12**, 888–892 (1987)

34. P.J. Mears, L. Reekie, I.M. Jauncey, D.N. Payne, Sintering behaviour of pressed transition of Er^{3+}. Electron. Lett. **23**, 1026–1031 (1987)

35. F. Auzel, P. Goldner, Towards rare-earth clustering control in doped glasses. Opt. Mater. **16**(1–2), 93–103 (2001)

36. M. Takesue, H. Hayashi, R. Smith Jr., Thermal and chemical methods for producing zinc silicate (willemite): a review. Prog. Cryst. Growth Charact. Mater. **55**(3–4), 98–124 (2009)

Chapter 2
Literature Review of Glass-Ceramic and Willemite Production from Waste Materials

2.1 Introduction

This chapter is divided into four major sections to provide a guideline and better understanding of the present work. The chapter begins with a general overview of glass and glass ceramics produced from waste (Sect. 2.2), followed by general information about willemite and willemite preparation methods in Sect. 2.3. Willemite preparation techniques are summarized in Sect. 2.4, and in Sect. 2.5, glass and glass-ceramic doping with rare earths (RE) is discussed, followed by an in-depth discussion on sintering in Sect. 2.6. Physical properties are discussed in Sect. 2.7, followed by microstructural properties and optical properties in Sects. 2.8 and 2.9, respectively.

2.2 Glass and Glass Ceramics Produced from Waste

It has to be accepted that there cannot be zero waste from any manufacturing, industrial, or energy conversion process, including power generation. It follows that for efficient use of the world's resources, recycling and reuse of waste are necessary. The versatility of the glass-ceramic production method has been shown by many researchers and involves using waste as raw materials, including coal fly ash [1–4], mud from zinc hydrometallurgy [5, 6], slag from steel production [7, 8], ash and slag from waste incinerators [9–13], red mud from alumina production [14], waste glass from lamp and other glass product manufacturing [15], and electric-arc furnace dust and foundry sands [16].

The first attempt to produce a glass ceramic by using waste material was reported as early as the 1960s and involved use of several types of slag of ferrous and

© The Author(s), under exclusive licence to Springer Nature Switzerland AG 2019   13
G. V. Sarrigani, I. S. Amiri, *Willemite-Based Glass Ceramic Doped by Different Percentage of Erbium Oxide and Sintered in Temperature of 500-1100C*,
SpringerBriefs in Electrical and Computer Engineering,
https://doi.org/10.1007/978-3-030-10644-7_2

nonferrous metallurgy, ash, and waste from the mining and chemical industries [17]. A secondary glass-ceramic manufacturing route—that of sintered glass ceramics—has been established since the 1960s [18].

To create a suitable parent glass for crystallization, additions to the waste are often required. Blast furnace slag was the first silicate waste to be thoroughly investigated as a source material for glass ceramics [19]. This slag consists of CaO, SiO_2, and MgO in decreasing amounts as the main constituents, together with minor constituents such as MnO, Fe_2O_3, and S. The first attempt to commercialized a glass ceramic from slag was made by the British Iron and Steel Research Association in the late 1960s [20]. This glass ceramic was known as Slagceram, and it was produced by a conventional, two-stage, heat treatment method [20].

Francis et al. [21, 22] explored the combination of coal fly ash and soda lime glass, using the powder route. In this case, the parent glasses were ground and then were given a sintering/crystallization treatment. Soda lime glass or soda lime silicate (SLS) glass is a well-recognized type of glass. It is usually use in windowpanes, glass containers (bottles and jars) for beverages and food, and some commodity items [21, 22]. It has been reported that the majority composition of SLS glass is silicon oxides (up to 75 wt.%), calcium oxides, and sodium oxides (SiO_2, CaO, and Na_2O) [18, 23–29]. Also there are minor percentages of several other oxides such as alumina and magnesia in SLS composition.

Willemite (Zn_2SiO_4), which is a good host for rare earths to be used in telecommunications, has been produced by different methods from pure material. However, there is a lack of research on willemite prepared by use of waste materials. This work discusses production of willemite using SLS glass waste material. This willemite has the benefit of using waste material as a main element. Nowadays, use of landfilling poses a major problem not only for large industrial countries but also for small cities. Use of waste materials and conversion of them into beneficial products is the best way to avoid filling up landfill sites.

2.3   Willemite and Methods Used for its Preparation

2.3.1   Willemite

The mineral well known to geologists by its inorganic name, willemite (Zn_2SiO_4), originates in nature. It was discovered in 1829 by Armand Lévy in Moresnet, a neutral territory that existed from 1816 to 1920 between the Netherlands and Prussia and is now known as La Calamine (Kelmis in German) in the Province of Liege, Belgium [30–32]. Willemite is named after Willem I of Orange-Nassau, King of the Netherlands, who reigned from 1815 to 1840.

Willemite has since been found to be distributed all over the Earth [33–37]. Willemite found in Franklin (NJ, USA) in the days of the Second Industrial Revolution was noted to exhibit green luminescence, which was attributed to the presence of manganese [38]. Artificial Franklin's willemite—Mn-doped zinc silicate (Zn_2SiO_4:Mn^{2+})—has been used as a phosphor in fluorescent lamps, neon discharge lamps, oscilloscopes, black-and-white televisions, color televisions, and many other displays and lighting devices since the 1930s [39–43]. Zn_2SiO_4:Mn^{2+} was specified as the P1 phosphor by the Electronics Industries Association (EIA) in 1945, and this symbol is widely used even today.

At present, Zn_2SiO_4:Mn^{2+} is used in large volumes in the most advanced televisions, plasma display panels (PDPs) because of its high luminescence efficiency, high color purity, and high chemical and thermal stabilities [39, 44, 45]. Zinc silicate, especially Zn_2SiO_4:Mn^{2+}, is one of the most widely applied and attractive materials that has appeared over the last 180 years, and it has been widely researched. Following its discovery, the focus on zinc silicate was directed toward its occurrence, crystallography, luminescence, and use as an industrial material. Bunting [46] confirmed the equilibrium phase in the ZnO–SiO_2 system, as shown in Fig. 2.1. Williamson and Glasser [47] discovered that the melting point of willemite was 1498 °C rather than 1512 °C (Fig. 2.1). Previously reported crystalline phases of willemite are presented in Table 2.1. Marumo and Syono [50] and Syono et al. [51] confirmed the crystallography of willemite II, III, and IV. Ringwood and Major [52] demonstrated Zn_2SiO_4-V sintering at 900 °C with 15 GPa pressure and sintering at 1000 °C. Doroshev et al. [53] studied Zn_2SiO_4-VI by decomposing $ZnSiO_3$ by sintering at 1100–1400 °C and 11–12.5 GPa pressure [46, 54].

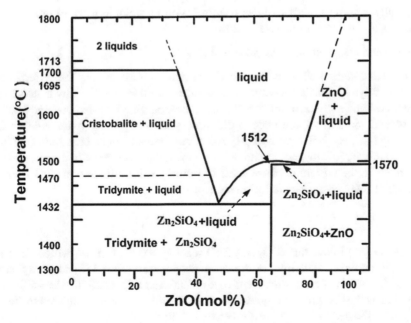

Fig. 2.1 Equilibrium phase in the ZnOeSiO₂ system, adapted from cristobalite and tridymite, which are polymorphs of crystalline SiO_2

Table 2.1 Synthetic conditions of willemite

Phase	Temperature (°C)	Pressure (GPa)	Si to Zn molar ratio	Reference
α(I)	800–1600	<3	0.5	[48]
β	700–800	Ambient	0.5–1.7	[47]
γ	1500–1600	Ambient	1	[49]
II	>800	8–10	0.5	[50]
III	>800	10–13	0.59	[51]
IV	>800	10–13	0.53	[51]
V	>900	>13	0.5	[52]
VI	1100–1400	11–12.5	0.5	[53]

Nowadays, different kinds of material are used to produce willemite; however, use of SLS glass waste material as a source of silicate is always a superior choice because this glass is inexpensive, has a low melting point (~700 °C), is chemically durable, and is relatively easy to melt and form. SLS glass was selected for this study because it has good chemical resistance, high stability, and high viscosity at a liquid temperature, and it allows compositional control of the index of refraction and the coefficient of thermal expansion [55].

2.3.2   Method Used for Willemite Preparation

Generally, different methods have been reported as suitable ways to produce willemite, and they are summarized below.

1. Conventional method (solid state and sintering)

The solid-state method is the most important process among the different elementary ways in which phosphors are produced. In this process, raw materials are well mixed and sintered at different temperatures ranging between 900 and 1500 °C for several hours in the presence of different gases that provide oxidizing (air, O_2), reducing (CO, H_2, NH_3), or inert (N_2, Ar) atmospheres in an electric furnace to produce an inorganic phosphor crystal with a dopant that is doped into the structure. In this method, the willemite phase starts to form by solid diffusion of ZnO from the surface of SiO_2, based on the following equation:

$$SiO_2 + 2ZnO = Zn_2SiO_4$$

Tammann [56] found that SiO_2 and ZnO react at 700 °C to form willemite. Pabst [57] also produced willemite at the same temperature. King [58] showed that the heat of formation of zinc silicate with use of ZnO and SiO_2 at 25 °C was −6.90 kcal/mol. The activation energy required to produce willemite was calculated by Takagi [59], who showed that the activation energy is 40.8 kcal/mol.

2. Sol–gel method

Generally, the sol–gel method is considered the proper way to prepare fine particles, improve the particle size distribution, or control the uniformity of the particle shape.

In this method a gel mixture is prepared by reaction of the mixture with a liquid such as ethanol or water. After that, the prepared gel is dried at a specific temperature, followed by calcination, to produce a fine crystalline phase. Morimo and Matae [60] studied preparation of Zn_2SiO_4:Mn^{2+} by the sol–gel method for lowering the processing temperature. Su and Johnson [61] used the sol–gel method to create a zinc compound $(Zn[OSi(O_tBu)_3]_2)_2$ that could be transformed into α-phase Zn_2SiO_4 by calcination at 1100 °C. Yang et al. [62, 63] suggested heating of the precursor prepared by using the sol–gel method in a microwave oven (with a frequency of 2450 MHz and power of 800 W) to obtain α-phase Zn_2SiO_4 instead of using calcination. Reynaud et al. [64] compared use of water conditions with zinc acetate and an aqueous $H_2Si_2O_2$ solution as raw materials and reported that α-phase Zn_2SiO_4 was formed after 6 h under hydrothermal conditions at 200 °C.

3. Hydrothermal and solvothermal method

Generally, hydrothermal conditions have the potential to produce materials at lower temperatures than those required in the solid-state method. In this method, two small copper tubes, in which an oxide and amorphous SiO_2 are separately loaded with water, are placed in a steel autoclave with water. Experimental conditions are achieved by heating of the autoclave. With use of the hydrothermal and solvothermal method, materials that have lower luminescence than materials synthesized with a solid-state reaction can be obtained. The fact that materials produced by the hydrothermal and solvothermal method have lower luminescence than materials synthesized with a solid-state reaction is due to the lower crystallinity of the material, owing to the short reaction time and low reaction temperature. Zhang et al. [65] produced α-phase Zn_2SiO_4 in water and ethanol over a 12-h reaction time at a temperature of 110 °C; however, the x-ray diffraction (XRD) peaks of the prepared material were very broad, indicating very low crystallinity.

4. Supercritical water method

In the supercritical water method, willemite is the most stable phase of the ZnO–SiO_2–H_2O system on a long scale. In this technique, sintered ZnO, SiO_2 blocks, and acid of the desired concentrations are sealed into a gold capsule or a fused silica tube. The gold capsule or fused silica tube is placed in a 5.5-cm^3 pressure vessel, which can be pressurized with water up to 2000 kg/cm^2 at 700 °C [66]. With use of this method a dense crystal of willemite can be produced within a very short reaction time. The supercritical water method may have the potential to be used in a practical process to create willemite with a low environmental burden. Komatsu et al. [67] reported production of α-phase Zn_2SiO_4 single crystals with prism shapes, which were several millimeters in length, in supercritical water conditions of 400 °C and 39 MPa, with a reaction time of 7 h.

5. Vapor method

Chemical vapor synthesis (CVS) is a method in which a solution including raw materials is vaporized and decomposed in a furnace at a high temperature. Roy et al. [68] studied and prepared a mixture of α- and β-phase willemite:Mn^{2+} with a particle size around 30 nm by chemical vapor synthesis with siloxy-substituted tetranuclear heterocubane $(MeZnOSiMe_3)_4$ as a raw material at a process temperature of 750–900 °C for 2–4 h.

2.4   Summary of Preparation Techniques

Generally, all of the aforementioned techniques have their own advantages and disadvantages; for instance, although the sol–gel method is considered the proper way to prepare fine particles, it needs a long reaction time and micropores are present after the synthesis. Also, the need for expensive autoclaves, safety issues during the reaction process, and the impossibility of observing the reaction process are some disadvantage of hydrothermal/solvothermal synthesis. With regard to the supercritical water technique, wide material development and research on supercritical water chemistry under radiation is needed. The poor step coverage, the difficulty of forming an alloy, and the lower throughput due to a low vacuum can be considered disadvantages of the vapor method. On the other hand, during the solid-state and sintering method, the pores and open channels that exist between grains of the compact tend to be removed, contributing to increased crystallinity of the powder and increased density, strength, toughness, and corrosion resistance of the material [68]. Also, it has been investigated that the microstructure of a glass ceramic prepared by solid-state sintering contains grains bonded together with a small amount of residual porosity [69].

2.5   Glass and Glass Ceramics Doped with Rare Earths

Glass ceramics doped with rare-earth ions have attracted significant interest because of their vast usage in the field of laser technology and optical communications. Various glass ceramics such as silicates and phosphates have been reported to be suitable hosts for rare earth; however, problems such as low chemical stability and a low transition temperature reduce the usage of phosphates as suitable hosts for rare earth.

Silicate glass ceramics have far superior chemical stability, which is an important factor in the ion exchange technique to produce optical wave guides. Recently, different rare earths have been studied as dopants for silicate glass and glass ceramics. Du and Kokou [70] studied the optical properties of Eu^{3+}-doped silica and sodium silicate glass ceramic. They found that europium ions tended to have less of a clustering tendency in sodium silicate glass. Pan et al. [71] investigated the effect of Tb^{3+} doping on lithium lanthanum aluminosilicate glass ceramic. It was found that terbium-activated lithium lanthanum aluminosilicate exhibited good ultraviolet (UV)–excited luminescence and radioluminescence. Brandily-Anne et al. [72] studied the influence of Ce^{3+} doping on the absorption spectra of multicomponent silicate glasses. It was reported that UV exposure induced a new band resulting from Ce^{3++} and electron color centers. However, there has been more interest in using trivalent erbium (Er^{3+})–doped silicate to obtain elements and source in telecommunication systems because the 1535-nm window is the wavelength region of choice for telecommunications and fortuitously coincides with the 1535-nm inter-4f $^4I_{13/2}$ → $^4I_{15/2}$ transition of the Er^{3+} ion. In 2004, Shih studied the effect of doping with trivalent Er^{3+} on sodium phosphate ($Na_2O–Er_2O_3–P_2O_5$) and found that it induced depolymerization of the glasses at the Q^3 tetrahedral sites, and the molar volume and T_g increased with increasing Er_2O_3 content. Chillcce et al. [73] studied the physical and optical properties of trivalent erbium–doped oxyfluoride tellurite glasses ($Er_2O_3–TeO_2–ZnF_2–ZnO$). They reported that the emission cross-sectional spectra of Er^{3+} ions in the oxyfluoride tellurite glasses with high concentrations of ZnF_2 present had narrow bandwidths that were very similar to those of fluoride glasses. The preparation and characterization of erbium-doped TeO_2-based oxyhalide glasses were studied by Fortes et al. [74]. The 1500-nm photoluminescence spectra and metastable level lifetimes of these glasses, especially with 3–5 mol.% $ErCl_3$, suggested that they may be suitable for optical amplifier applications. Courrol et al. [75] showed that the fluorescence lifetime was 0.40 ms, and better mechanical resistance under a high-brightness diode laser was observed. They also reported that glass doped with 0.10 wt.% Er_2O_3 showed interesting spectroscopic features suitable for laser action at 1523 nm. The luminescence of Er^{3+}-doped bismuth borate glasses was studied by Oprea et al. [76]. They reported that Er^{3+}-doped bismuth borate could be an interesting material for a large range of optical applications such as fiber amplifiers or lasers. Also, Rada et al. [77] studied the spectroscopic features and ab initio calculations on the structure of Er^{3+}-doped zinc borate glasses and glass ceramics. They concluded that the introduction of erbium oxide into the zinc borate matrix would increase the BO_4 structural units and form an $ErBO_3$ crystalline phase, which would improve the optical behavior. By studying the ab initio results they found that the Zn^{2+} ions acted as modifiers and generated bonding defects by breakage of the B–O–B bonds. Bento dos Santos et al. [78] investigated the spectroscopic and optical properties of Er^{3+}-doped germane silicate glasses and glass ceramic. They reported that the absorption and emission spectra increased in intensity with the Er^{3+} concentration, while the full width at half maximum increased from 40 to 47 nm between 0.1% and 4% $ErO_{1.5}$.

It has been shown that transparent glass ceramic may be a reliable alternative system to control the chemical parameters of rare earth and thus may avoid unfavorable effects such as clustering. In fact, glass-ceramic materials are of great importance in the area of photonics, because they mix the optical and mechanical properties of glass with those of crystal-like rare-earth ions. Moreover, it has been confirmed that in the presence of rare-earth ions, glass ceramics have better optical properties than the starting glass [79]. Different rare earths have been used to dope different classes of glass ceramics. Generally, there are five major rare-earth candidates for doping glass ceramics for use in optical communications: Tm^{3+}, Dy^{3+}, Pr^{3+}, Er^{3+}, and Nd^{3+}. Among the rare-earth ions, Er^{3+} is of greatest interest because of its emission near 1535 nm, corresponding to the third telecommunication window. Different kinds of glass ceramics have been used as hosts for Er^{3+}. Jestin et al. [80] used Er^{3+} to activate HfO_2-based glass-ceramic wave guides for use in photonics, and Bento dos Santos et al. [78] utilized Er^{3+} ions as a dopant in niobium germane silicate (40 GeO_2–10 SiO_2–25 Nb_2O_5–25 K_2O) glass ceramics.

2.6   Sintering

Generally, raw materials need a high melting point and, because of that, the creation of glass ceramics includes a heat treatment stage in which a powder, already formed into a special form, is converted into a dense solid. This step is referred to as the firing or sintering process. In general, to produce a material with particular properties, the required microstructure and a process to create this microstructure are needed. The main purposes of sintering studies are to gain knowledge of the particle size, temperature, particle packing, composition, and microstructure.

2.6.1   Types of Sintering Process

There are four main types of sintering. The type of sintering is generally dependent on the composition being sintered and, to some extent, which second phase is formed during the sintering.

1. Solid-state sintering: In this type of sintering the already-formed green body is heated to a temperature that is generally 0.5–0.9 of the melting point. In general, there is no liquid in solid-state sintering; therefore, atomic diffusion causes joining of particles and reduction of porosity.
2. Liquid-phase sintering: There is a small amount of liquid—less than a few percent of the original solid mixture—involved in this type of sintering. In fact, the amount of liquid present is not enough to fill the pore space, so additional processes are needed to achieve full densification.
3. Vitrification: In this process a dense product can be obtained by creation of a liquid, flow of the liquid into pores, and crystallization or vitrification during the cooling of the liquid. In fact, this is due to the sufficiency of a large volume of water formed through heating to fill the volume of the remaining pores.
4. Viscous sintering: This kind of sintering occurs by viscous flow of glass under the effect of surface tension. In this process a consolidated mass of glass particles is fired near to, or above, its softening temperature.

2.6.2 Steps in the Sintering Process

The sintering process can be divided into three different stages specified according to the period of the microstructure, in order to force correspondence among simple, established sintering patterns.

The first stage starts as soon as some degree of atomic mobility is gained and, during this stage, among the individual particles, a sharply concave is formed. At this stage the amount of densification is small—usually the first 5% of linear shrinkage—and it may be remarkably low because of high activity of coarsening mechanisms.

At the second stage the high inflections of the first stage have been balanced, so the microstructure includes a three-dimensional structure of solid particles and persistent, channel-like pores. In fact, the porosity at the second stage is estimated to be 5–10%; thus, this stage covers most of the densification, and the grain increase (coarsening) starts to become considerable.

During the final stage, the grain increase can be faster and, as a result, the removal of the last few percent of porosity is very difficult. This is because of the breaking down of the channel-like pores into isolated and closed voids.

2.6.3 Mechanism of Sintering

Basically, sintering of crystalline material can happen by some major mechanisms such as evaporation or vapor transport, surface diffusion, lattice diffusion, an increase in the grain boundary, and dislocation motion. Figure 2.2 depicts a schematic diagram of the matter paths for two sintering particles. There is a distinction between densifying and nondensifying mechanisms. Vapor, surface, and lattice diffusion from the particle surface causes a neck increase and coarsening of particles without densification. Generally, grain boundary and lattice diffusion from the grain boundary to the neck are the most important densifying mechanisms in polycrystalline ceramics because diffusion from the grain boundary to the pores allows a neck increase, as well as shrinkage. Plastic flow by dislocation motion can cause neck growth and shrinkage through deformation of the particles in response to the sintering stress.

2.7 Physical Properties

Sintering is the term used to describe consolidation of a product during firing. Consolidation implies that within the product, particles have joined together into a strong aggregate. The term sintering is often interpreted as implying that shrinkage and densification have occurred [81]. The process of firing glass powder to create a

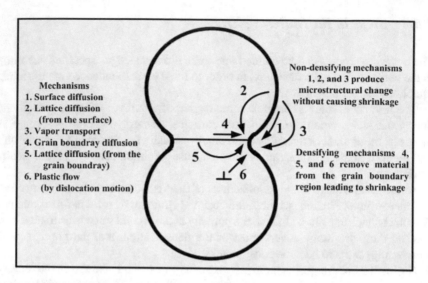

Fig. 2.2 Sintering mechanisms in a system of two particles

glass ceramic is called solid-state sintering. Generally, the process of solid-state sintering affects the glass-ceramic density, porosity, mechanical properties, and microstructural evolution. Higher sintering temperatures will cause the glass-ceramic particles to be condensed. This tends to reduce the porosity and enhance the density, as well as the grain size, of the glass ceramic. In general, an increase in the sintering temperature results in greater density and grain growth, which decreases the porosity and the number of grain boundaries [82]. Thus, it can be indicated that, generally, the porosity exhibits a reverse trend to that of the linear shrinkage, and the bulk density exhibits similar behavior to that of the linear shrinkage [83].

2.8   Microstructural Characteristics

There are three steps in solid-phase sintering that occur in the microstructural evolution. In general, the steps in solid-phase sintering rely on the sintering temperature and time, the particle size, and the nature and shape of the material. At the first stage of sintering, when all of the particles come into contact, only a small mass of particles is involved in creating the neck, so the neck increase of the particles and the decrease in porosity are insignificant in the microstructure. However, at the second stage—the intermediate stage—sintering takes place as a large mass of particles become involved in the neck increase. As a result, in this step, considerable densification can be seen in the microstructure. At the final stage, the increase in grain size and densification are evident. In fact, some grains are joined together and larger-sized grains are produced. As a result, the average grain size increases and fewer grains can be observed in a unit area of a micrograph. With regard to willemite, which has high percentages of ZnO and SiO_4, the glass-ceramic network may consist of Zn–O, ZnO_4, and Si–O–Si bonds. In different studies, different locations of the willemite band have been reported. Tsai et al. [84], Marumo and Syono [50], Takesue et al. [85], and Rooksby and McKeag [49] characterized willemite using XRD and Fourier transform infrared spectroscopy (FTIR). The diffraction peaks of willemite (Zn_2SiO_4) at $2\theta = 12.568, 21.928, 25.428, 31.388, 33.868, 38.688, 44.888, 48.788, 57.468, 60.688, 65.448$, and 70.108 could be indexed as (1 1 0), (3 0 0), (2 2 0), (1 1 3), (4 1 0), (2 2 3), (6 0 0), (3 3 3), (7 1 0), (6 3 0), (7 1 3), and (4 1 6), respectively. They also reported the FTIR spectra of Zn_2SiO_4 powder. Vibrational modes typical of this material were observed—866 cm^{-1} (ν_1 SiO_4); 901, 931, and 977 cm^{-1} (ν_3 SiO_4); 460 cm^{-1} (ν_4 SiO_4); 573 cm^{-1} (ν_1 ZnO_4); and 613 cm^{-1} (ν_3 ZnO4)—where ν_1 stands for totally symmetric stretching, ν_3 is asymmetric stretching, and ν_4 is asymmetric deformation [86–90]. The appearance of the vibrations of the SiO_4 and ZnO_4 groups clearly suggested formation of the Zn_2SiO_4 phase.

2.9 Optical Properties

The incorporation of rare-earth ions into a variety of glasses has been considered an important factor in the development of optical devices based on active glass materials—such as infrared (IR) lasers, IR-to-visible up-converters, fiber, and wave guide amplifiers for optical transmission networks—in the last decade [91]. Rare earths (especially the most important one, Er^{3+}), which are optically active ions, are usually incorporated into the silicate crystalline phase. Several advantages of mixed silicate glass ceramics have been observed. In fact, they are characterized by the excellent optical properties of rare-earth ions in silicate, with the high chemical stability and mechanical properties of the oxide matrix. The good properties of silicate glass ceramics need to be optimized via high-temperature heat treatment. For these reasons, systematic study of both the thermal and optical properties of silicate glass ceramics with high-temperature heat treatment is needed—for example, to understand the fundamental characteristics of the glass and to optimize both the crystallization temperature and the photoluminescence intensity. Different studies have analyzed the optical properties of Er^{3+} in silicate glass ceramics. Bento dos Santos et al. studied the optical and spectroscopic properties of Er-doped niobium germanosilicate glass ceramics. In the UV-VIS characterization, they reported that peaks located at 980, 820, 650, 520, and 490 nm could be assigned to states from the $^4I_{15/2}$ state to $^4I_{11/2}$, $^4I_{9/2}$, $^4F_{9/2}$, $^4S_{3/2}$, and $^2H_{11/2}$. Polman [92], Laegsgaard [93], and Auzel et al. [94] studied the role of Er^{3+} ions used as a dopant in glass-ceramic hosts. They reported that there was a strong tendency for erbium ions to cluster together in precipitates that provided further nonradiative de-excitation pathways and that, for this reason, only relatively modest concentrations of erbium could be incorporated into silicon. Much of the work in this area has therefore focused on overcoming these problems.

References

1. A.R. Boccaccini, M. Bücker, J. Bossert, Glass and glass-ceramics from coal fly ash and waste glass. Tile Brick Int. **12**, 515–518 (1996)
2. M. Erol, S. Küçükbayrak, A. Ersoy-Meriçboyu, M.L. Öveçoğlu, Crystallization behaviour of glasses produced from fly ash. J. Eur. Ceram. Soc. **21**(16), 2835–2841 (2001)
3. S. Kumar, K.K. Singh, P. Ramachandrarao, Synthesis of cordierite from fly ash and its refractory properties. J. Mater. Sci. Lett. **19**(14), 1263–1265 (2000)
4. R. Siddique, Wood ash, in *Waste Materials and By-Products in Concrete*, (Springer, London, 2008), pp. 303–321
5. A. Karamanov, P. Pisciella, C. Cantalini, M. Pelino, Influence of Fe^{3+}/Fe^{2+} ratio on the crystallisation of iron-rich glasses made with industrial wastes. J. Am. Ceram. Soc. **83**, 3153–3157 (2000)
6. L. Montanaro, N. Bianchini, J.M. Rincon, M. Romero, Sintering behaviour of pressed red mud wastes from zinc hydrometallurgy. Ceram. Int. **27**(1), 29–37 (2001)
7. E. Fidancevska, B. Mangutova, D. Milosevski, M. Milosevski, J. Bossert, Obtaining of dense and highly porous ceramic materials from metallurgical slag. Sci. Sinter. **35**, 85–91 (2003)
8. G.A. Khater, The use of Saudi slag for the production of glass-ceramic materials. Ceram. Int. **28**(1), 59–67 (2002)
9. F. Andreola, L. Barbieri, A. Corradi, I. Lancellotti, New marketable products from inorganic residues. Am. Ceram. Soc. Bull. **3**, 9401–9408 (2004)
10. L. Barbieri, A. Corradi, I. Lancellotti, Thermal and chemical behaviour of different glasses containing steel fly ash and their transformation into glass-ceramics. J. Eur. Ceram. Soc. **22**(11), 1759–1765 (2002)
11. H.S. Kim, J.M. Kim, T. Oshikawa, K. Ikeda, Production and properties of glass-ceramics from incinerator fly ash. Mater. Sci. Forum **439**, 180–185 (2003)
12. J. Kim, H. Kim, Glass-ceramic produced from a municipal waste incinerator fly ash with high Cl content. J. Eur. Ceram. Soc. **24**(8), 2373–2382 (2004)
13. L. Stoch, Homogeneity and crystallisation of vitrified municipal waste incinerator ashes. Soc. Glass Technol. **45**, 71–73 (2004)
14. P. Zhang, J. Yan, Mossbauer and infrared spectroscopy investigation on glass-ceramics using red mud. Z. Metallk. **91**, 764–768 (2000)
15. M.M. Sokolova, V.S. Perunova, V.V. Sepanov, N.V. Kozlov, A glass ceramic material based on the waste from lamp production. Glas. Ceram. **43**, 133–135 (1986)
16. Z. Gao, C.H. Drummond III, Thermal analysis of nucleation and growth of crystalline phases in vitrified industrial wastes. J. Am. Ceram. Soc. **82**, 561–565 (1999)
17. W. Holand, G. Beall, Thermal expansion properties of a spodumene-willemite glass ceramic, in *Glass-Ceramic Technology*, (The American Ceramic Society, Westerville, 2002)
18. E. Bernardo, R. Castellan, S. Hreglich, Sintered glass-ceramics from mixtures of wastes. Ceram. Int. **33**(1), 27–33 (2007)
19. R.D. Rawlings, J.P. Wu, A.R. Boccaccini, Glass-ceramics: their production from wastes-a review. J. Mater. Sci. **41**(3), 733–761 (2006)
20. M.W. Davies, B. Kerrison, W.E. Gross, W.J. Robson, D.F. Wichell, Slagceram: a glass-ceramic from blast-furnace slag. J. Iron Steel Inst. **208**, 348–370 (1970)
21. A.A. Francis, A.R. Boccaccini, R.D. Rawlings, Production of glass-ceramics from coal ash and waste glass mixtures. Key Eng. Mater. **206-213**, 2049–2052 (2002)
22. A.A. Francis, R.D. Rawlings, R. Sweeney, A.R. Boccaccini, Crystallization kinetic of glass particles prepared from a mixture of coal ash and soda-lime cullet glass. J. Non-Cryst. Solids **333**(2), 187–193 (2004)
23. L. Barbieri, A. Corradi, I. Lancellotti, G.C. Pellacani, Sintering and crystallisation behaviour of glass frits made from silicate waste. Glass Technol. **44**, 184–190 (2003)

24. E. Bernardo, M. Varrasso, F. Cadamuro, S. Hreglich, Vitrification of wastes and preparation of chemically stable sintered glass-ceramic products. J. Non-Cryst. Solids 352(38–39), 4017–4023 (2006)
25. A. Karamanov, Granite like materials from hazardous wastes obtained by sinter crystallisation of glass frits. Adv. Appl. Ceram. 108, 14–21 (2009)
26. N. Marinoni, D. D'Alessio, V. Diella, A. Pavese, F. Francescon, Effects of soda–lime–silica waste glass on mullite formation kinetics and micro-structures development in vitreous ceramics. J. Environ. Manag. 124, 100–107 (2013)
27. S.N. Salama, S.M. Salman, H. Darwish, The effect of nucleation catalysts on crystallization characteristics of aluminosilicate glasses. Ceramics-Silikáty 46, 15–23 (2002)
28. S.R. Scholes, *Modern Glass Practice* (CBI Publishing Company, Boston, 1975), pp. 1–493
29. T. Toya, Y. Kameshima, A. Yasumori, K. Okada, Preparation and properties of glass-ceramics from wastes (Kira) of silica sand and kaolin clay refining. J. Eur. Ceram. Soc. 24(8), 2367–2372 (2004)
30. M. Braun, Z.D. Geolog, Thermal and chemical methods for producing zinc silicate (willemite): A review. Gesellschaft 9, 354–370 (1857)
31. M. Le'vy, M. Annales, Dont l'auteur est très ... un silicate de zinc nouvellement découvert à la Vieille-Montagne, célèbre mine de zinc, Paris. 4e`me se´rie 4, 507–520 (1843)
32. J. Schneider, M. Boni, C. Laukamp, T. Bechstädt, V. Petzel, Willemite (Zn$_2$SiO$_4$) as a possible Rb–Sr geochronometer for dating nonsulfide Zn–Pb mineralization: examples from the Otavi Mountainland (Namibia). Ore Geol. Rev. 33(2), 152–167 (2008)
33. M. Boni, D. Large, Willemite in the Belgian non-sulphide deposits: a fluid inclusion study. Econ. Geol. 98, 715–729 (2003)
34. P. Bowen, C. Carry, From powders to sintered pieces: forming, transformations and sintering of nanostructured ceramic oxides. Powder Technol. 128(2–3), 248–255 (2002)
35. V. Coppola, M. Boni, H.A. Gilg, G. Balassone, L. Dejonghe, The "calamine" nonsulfide Zn–Pb deposits of Belgium: petrographical, mineralogical and geochemical characterization. Ore Geol. Rev. 33(2), 187–210 (2008)
36. M.W. Hitzman, N.A. Reynolds, D.F. Sangster, C.R. Allen, C.E. Carman, Preparation and characterizations of green phosphors. Econ. Geol. 98, 685–714 (2003)
37. F.H. Pough, Production and properties of zinc silicate mineral. Am. Mineral. 26, 92–102 (1941)
38. C. Palache, T. Feldmann, J. Stel, C.R. Ronda, P.J. Schmidt, Sintering effects on mechanical properties of glass-reinforced zinc silicate. Am. Mineral. 13, 330–333 (1928)
39. C. Feldmann, T. Justel, C.R. Ronda, P.J. Schmidt, Tripolyphosphate as precursor for REPO(4):Eu (3+) (RE = Y, La, Gd) by a polymeric method. Adv. Funct. Mater. 13, 511–516 (2003)
40. D.E. Harrison, Preparation and characterizations of Zn$_2$:Mn phosphors. J. Electrochem. Soc. 107, 210–217 (1960)
41. H.W. Leverenz, *An Introduction to Luminescence of Solids* (Wiley, New York, 1950), pp. 399–401
42. T. Minami, Erbium-doped glasses for fiber amplifiers. Solid State Electron. 47, 2237–2243 (2003)
43. C.R. Ronda, Characterisation of a glass and a glass-ceramic obtained from municipal incinerator ash. J. Lumin. 72-74, 49–54 (1997)
44. H. Liang, Q. Zeng, Z. Tian, H. Lin, Q. Su, G. Zhang, Y. Fu, Intense emission of Ca5(PO$_4$)3F: Tb^{3+}under VUV excitation and its potential application in PDPs. J. Electrochem. Soc. 154, J177–J180 (2007)
45. S. Zhang, Structure and luminescence properties of Mn-doped Zn2SiO4 prepared with extracted mesoporous silica. Mater. Res. Bull. 46(6), 791–795 (2006)
46. E.N. Bunting, Phase equilibria in the system SiO$_2$-ZnO. Bur. Standards J. Res 4, 131–136 (1930)

47. J. Williamson, F.P. Glasser, Optical and physical properties of Er^{3+}-doped oxy-fluoride tellurite glasses. Glas. Phys. Chem. **5**, 52–59 (1964)
48. Y. Syono, S. Akimoto, Y. Matsui, Crystallization kinetic of glass particles prepared from a mixture of coal ash and soda-lime cullet glass. J. Solid State Chem. **3**, 369–380 (1971)
49. H.P. Rooksby, A.H. McKeag, Trans. Faraday Soc. **37**, 308–311 (1941)
50. F. Marumo, Y. Syono, Effects of soda–lime–silica waste glass on transition of Er^{3+} formation kinetics and micro-structures development in vitreous ceramics. Acta Crystallogr. B **27**, 1868–1870 (1971)
51. Y. Syono, S. Akimoto, Y. Matsui, High pressure transformations in zinc silicates. J. Solid State Chem. **3**, 369–380 (1971)
52. A.E. Ringwood, A. Major, High pressure transformations in zinc germanates and silicates. Nature **215**, 1367–1368 (1967)
53. A.M. Doroshev, M. Olesch, V.M. Logvinov, I.J. Malinovsky, Preparation and characterization of Er^{3+}-doped TeO_2-based oxyhalide glasses. Mineral **27**, 277–288 (1983)
54. E.N. Bunting, Synthesis, properties and mineralogy of important inorganic materials. J. Am. Ceram. Soc. **13**, 5–10 (1930)
55. B.G. Bagley, E.M. Vogel, W.G. French, G.A. Pasteur, J.N. Gan, J. Tauc, The optical properties of a soda-lime-silica glass in the region from 0.006 to 22 eV. J. Non-Cryst. Solids **22**(2), 423–429 (1976)
56. G. Tammann, Chemische Reaktionen in pulverförmigen Gemengen zweier Kristallarten. Z. Anorg. Allg. Chem. **149**(1), 21–98 (1925)
57. A. Pabst, Röntgenuntersuchung über die Bildung von Zinksilicaten. Z. Phys. Chem. **142A**(1), 227–232 (1929)
58. E.J. King, The phosphatases, alkaline phosphatase. Postgrad. Med. J. **27**(304), 64–66 (1951)
59. S. Takagi, Dynamical theory of diffraction applicable to crystals with any kind of small distortion. Acta Crystallogr. **15**(12), 1311–1312 (1962)
60. R. Morimo, K. Matae, Preparation of Zn_2SiO_4:Mn phosphors by alkoxide method. Mater. Res. Bull. **24**(2), 175–179 (1989)
61. H. Su, D.L. Johnson, Master sintering curve: a practical approach to sintering. J. Am. Ceram. Soc. **79**(12), 3211–3217 (1996)
62. H. Yang, J. Shi, M. Gong, A novel approach for preparation of Zn2SiO4: Tb nanoparticles by sol-gel-microwave heating. J. Mater. Sci. **40**(22), 6007–6010 (2005). https://doi.org/10.1007/s10853-005-2632-1
63. H. Yang et al., Synthesis and photoluminescence of Eu3+- or Tb 3+-doped Mg_2SiO_4 nanoparticles prepared by a combined novel approach. J. Lumin. **118**(2), 257–264 (2006)
64. L. Reynaud et al., A new solution route to silicates. Part 3: Aqueous sol-gel synthesis of willemite and potassium antimony silicate. Mater. Res. Bull. **31**(9), 1133–1139 (1996)
65. S. Zhang et al., Synthesis and electrochemical properties of Zn_2SiO_4 nano/mesorods. Mater. Lett. **100**, 89–92 (2013)
66. K. Kodaira, S. Ito, T. Matsushita, Hydrothermal growth of willemite single crystals in acidic solutions. J. Cryst. Growth **29**(1), 123–124 (1975)
67. K.-I. Komatsu, M. Mizuno, R. Kodaira, Effect of methionine on cephalosporin C and penicillin N production by a mutant of *Cephalosporium acremonium*. J. Antibiot. **28**(11), 881–888 (1975)
68. A. Roy, S. Polarz, S. Rabe, B. Rellinghaus, H. Zahres, F.E. Kruis, M. Driess, First preparation of nanocrystalline zinc silicate by chemical vapor synthesis using an organometallic single-source precursor. Chemistry **10**(6), 1565–1575 (2004)
69. L. Hench, J. Wilson, An introduction to bioceramics. Adv. Ser. Ceram. **18**, 1–389 (1999)
70. J. Du, L. Kokou, Europium environment and clustering in europium doped silica and sodium silicate glasses. J. Non-Cryst. Solids **357**(11), 2235–2240 (2011)
71. Z. Pan et al., Terbium-activated lithium–lanthanum–aluminosilicate oxyfluoride scintillating glass and glass-ceramic. Nucl. Instrum. Methods Phys. Res. Section A **594**(2), 215–219 (2008)

72. M.-L. Brandily-Anne et al., Specific absorption spectra of cerium in multicomponent silicate glasses. J. Non-Cryst. Solids **356**(44), 2337–2343 (2010)

73. E.F. Chillcce et al., Optical and physical properties of Er3+-doped oxy-fluoride tellurite glasses. Opt. Mater. **33**(3), 389–396 (2011)

74. L. Fortes et al., Preparation and characterization of Er^{3+}-doped TeO_2-based oxyhalide glasses. J. Non-Cryst. Solids **324**(1), 150–158 (2003)

75. L.C. Courrol et al., Lead fluoroborate glasses doped with Nd^{3+}. J. Lumin. **102–103**, 101–105 (2003)

76. I.-I. Oprea, H. Hesse, K. Betzler, Luminescence of erbium-doped bismuth–borate glasses. Opt. Mater. **28**(10), 1136–1142 (2006)

77. S. Rada et al., Spectroscopic properties and ab initio calculations on the structure of erbium–zinc-borate glasses and glass ceramics. J. Non-Cryst. Solids **358**(1), 30–35 (2012)

78. D. Bento dos Santos et al., Itaquaquecetuba formation palynostratigraphy, São Paulo Basin, Brazil. Rev. Bras. Paleontolog. **13**, 205–220 (2010)

79. M. Mortier, P. Goldner, C. Chateau, M. Genotelle, Erbium doped glass–ceramics: concentration effect on crystal structure and energy transfer between active ions. J. Alloys Compd. **323–324**, 245–249 (2001)

80. Y. Jestin et al., Erbium activated HfO_2 based glass–ceramics waveguides for photonics. J. Non-Cryst. Solids **20**, 494–497 (2007)

81. F. Oktar, G. Göller, Sintering effects on mechanical properties of glass-reinforced hydroxyapatite composites. Ceram. Int. **28**(6), 617–621 (2002)

82. A. Verma, R. Chatterjee, Effect of zinc concentration on the structural, electrical and magnetic properties of mixed Mn–Zn and Ni–Zn ferrites synthesized by the citrate precursor technique. J. Magn. Magn. Mater. **306**, 313–319 (2006)

83. N.Y. Mostafa, A.A. Shaltout, S. Abdel-Aal, A. El-maghraby, Sintering mechanism of blast furnace slag–kaolin ceramics. Mater. Des. **31**(8), 3677–3682 (2010)

84. M.-T. Tsai et al., Photoluminescence of titanium-doped zinc orthosilicate phosphor gel films. IOP Conf. Ser. Mater. Sci. Eng. **18**(3), 032012 (2011)

85. M. Takesue, H. Hayashi, R.L. Smith, Thermal and chemical methods for producing zinc silicate (willemite): a review. Prog. Cryst. Growth Charact. Mater. **55**(3), 98–124 (2009)

86. G.T. Chandrappa, S. Ghosh, K.C. Patil, Synthesis of glass-ceramic. J. Mater. Synth. Process. **7**(1), 273–282 (1999)

87. C. Lin, P. Shen, Sol-gel synthesis of zinc orthosilicate. J. Non-Cryst. Solids **171**(3), 281–289 (1994)

88. R.P. Sreekanth Chakradhar, B.M. Nagabhushana, G.T. Chandrappa, K.P. Ramesh, J.L. Rao, J. Opt. Soc. Am. **121**, 10250–10259 (2004)

89. P. Taret, Etude infra-rouge des orthosilicates et des orthogermanates Une nouvelle methode d'interprétation des spectres. Spectrochim. Acta **18**(4), 467–483 (1962)

90. M. Bosca, L. Pop, G. Borodi, P. Pascuta, E. Culea, XRD and FTIR structural investigations of erbium-doped bismuth–lead–silver glasses and glass ceramics. J. Alloys Compd. **479**(1–2), 579–582 (2009)

91. I. Jlassi, H. Elhouichet, S. Hraiech, M. Ferid, Effect of heat treatment on the structural and optical properties of tellurite glasses doped erbium. J. Lumin. **132**(3), 832–840 (2012)

92. A. Polman, Erbium implanted thin film photonic materials. J. Appl. Phys. **82**(1), 1–39 (1997)

93. J. Laegsgaard, Theory of Al_2O_3 incorporation in SiO_2. Phys. Rev. B **65**(17), 174104 (2002)

94. F. Auzel, On the maximum splitting of the (2F7/2) ground state in Yb3+-doped solid state laser materials. J. Lumin. **93**(2), 129–135 (2001)

Chapter 3
Methodology for Preparation Samples from Waste and Techniques for Characterization

3.1 Introduction

This chapter explains the procedure of willemite synthesis using SLS glass waste material and pure ZnO then doped with pure Er_2O_3. In this work, different contents of trivalent erbium ranging from 1 to 5 wt.% were doped on willemite glass-ceramic and then sintered at 500–1100 °C. Thirty-six samples were prepared in producing willemite. Figure 3.1 illustrates the stages of producing willemite material. The study was continued with the analysis of structure, optical, and physical properties of erbium-doped willemite-based glass-ceramic.

3.2 Sample Preparation

Thirty-six series of samples were prepared by mixing 50 wt.% pure ZnO (99.99%, Aldrich) and 50 wt.% SLS glass powder (waste glass bottle). The KFC bottle from Taman Sri Serdang was used as the source for SLS glass material. To produce the SLS glass powder, the waste SLS glass bottle was crushed using mortar and pestle (in this technique most of the glass sample never contact with the sides of the mortar, so the content of contamination was very slight) followed by grounding to the size of <63 μm. By using this technique, the glass sample did not contact with the sides of the mortar. As a result the percentage of contamination was very slight.

© The Author(s), under exclusive licence to Springer Nature Switzerland AG 2019
G. V. Sarrigani, I. S. Amiri, *Willemite-Based Glass Ceramic Doped by Different Percentage of Erbium Oxide and Sintered in Temperature of 500-1100C*, SpringerBriefs in Electrical and Computer Engineering, https://doi.org/10.1007/978-3-030-10644-7_3

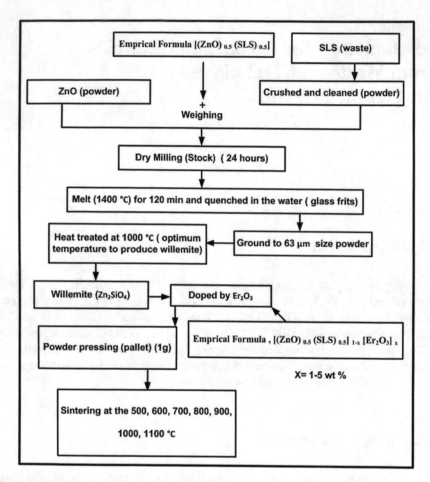

Fig. 3.1 The process of preparing raw material, willemite, and Er^{3+}-doped willemite

3.2.1   Weighing, Mixing, and Milling

An electronic digital weighing machine was used to weight SLS glass and ZnO powder. The electronic digital weighing has an accuracy of ±0.001 g. To obtain 100 g sample, 50 g SLS glass and 50 g ZnO were weighed and mixed thoroughly. After that the mixture was put to a milling container along with some milling ball in different sizes. The dry milling is recognized as a proper method to obtain fine and homogenous glass powder. Then, the milling container was brought to the milling machine and the batch of mixture was milled for 24 h.

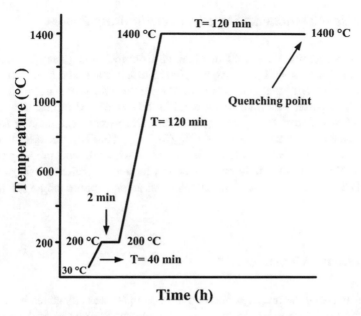

Fig. 3.2 Sample melt-quenching process

3.2.2 Glass Melting Method

The first thing to be investigated is the melting point when selecting the glass melt-ing procedures to be used. Generally inorganic glasses are melted at least 100 °C above the liquids' temperature to gain a viscosity that is low enough to allow homogenization and refinement to be achieved. Commonly, glass has different melting temperatures ranging from 750 to 2000 °C. The melting point is depending on the oxide composition of the materials. However in this work the temperature of 1400 °C was used as a melting point to melt the sample.

After the ball milling stage, the obtained mixture was shed to an alumina cruci-ble. The profile of melting process is shown in Fig. 3.2

3.2.3 Glass-Quenching Technique

The process of quenching melted glass mixture powder was done by rapidly shed-ding the melted powder from the furnace in water to form glass frits. After that the prepared glass frits were dried in air for 1 day and then were crushed into powder and sieve to the size of 63 μm.

3.2.4   Heat Treatment, Doping, and Sintering Process

To prepare willemite, the glass frits obtained from melt and quenching stage were ground to the size of <63 µm, then transferred to a furnace, and heated at 1000 °C sintering temperature that is an optimum temperature for producing willemite [1]. After that, prepared willemite was ground to the size of <63 µm using mortar and pestle and then doped with pure Er_2O_3 in different percentages ranging from 1 to 5 wt.%; the empirical formula of $[(ZnO)_{0.5}(SLS)_{0.5}]_{1-x}[Er_2O_3]_x$ was used. In order to dope the mentioned content of Er_2O_3, it was mixed with willemite properly followed by 24 h ball milling to gain homogenous powder and by pelleting using PVA as a binder. The sintering process of the samples was at temperature ranging from 500 to 1100 °C.

3.3   Density Measurement

The density (ρ) of the prepared glass-ceramic samples was calculated by a densimeter model MD-300S using Archimedes principle applying water as the immersion fluid. At the first stage, the samples were weighed in air, W_{air}, then in an immersion liquid (water), W_{water}, with the following density: $\rho_{water} = 1$ g cm³. The density of the samples was then measured using the following relationship (Eq. (3.1) [2]).

$$\rho = \frac{W_{air} \; \rho_{water}}{W_{air} - W_{water}}$$ (3.1)

The linear shrinkage of samples was calculated according to this relation (3.2) [3].

$$\%\text{Linear Shrinkage} = \frac{L_A - L_B}{L_A} \times 100$$ (3.2)

where *LA* is the symbol of the green body dimension and *LB* is the dry dimension of the sintered sample.

3.4   X-Ray Diffraction

A beneficial technique to investigate structure of material is X-ray diffraction in which various crystalline phases can be identified. In fact XRD patterns can specify whether a sample is crystalline or amorphous material. Actually crystals have a regularly repeating shape that is capable to diffract radiation of wavelength similar to the interatomic distances [4]. Figure 3.3 shows the schematic diagram of

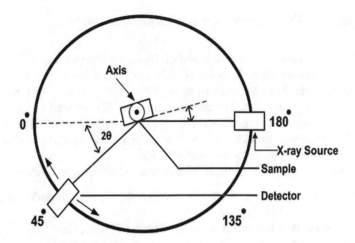

Fig. 3.3 Schematic diagram of X-ray diffractometer

Fig. 3.4 Bragg's law of diffraction

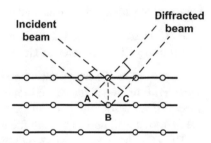

XRD. Figure 3.3 shows that in the first stage, a sample is lighten by a beam radiated from an X-ray source. After that a rotatory detector counted the X-ray at a 2θ angle, and a beam that came out of the sample lattice created patterns based on Bragg's law of diffraction that is displayed in Fig. 3.4.

The path between two waves undergoing constructive interference is presented by $2d \sin \theta$, where θ refers to the scattering angle. In fact Bragg's angle explains the conditions for constructive interference from successive crystallographic planes (h, k, l) of the crystalline lattice:

$$2d\sin\theta = n\lambda \qquad (3.3)$$

where n refers to an integer determined by the order and λ is the wavelength. A diffraction pattern is gained by measuring the intensity of scattered waves as a function of scattering angle.

The sample can be analyzed by XRD in the form of powder, pallet, or even thin film. However, in this study, all the glass-ceramic samples were analyzed in the form of powder. Samples were prepared by grinding using mortar and pestle to have homogenous sample. Then the powdered sample was put into the holder dedicated for powder form.

3.5   Fourier-Transform Infrared Spectroscopy

Infrared spectroscopy is the investigation of the action and reaction of infrared light with article. Infrared spectrum, a plot of estimated infrared spectrum against wavenumber of light, is the basic measurement gained from infrared spectroscopy. By the help of a machine called infrared spectrometer, the infrared spectrum can be obtained. Generally, there are different kinds of spectrometer around the world suitable to obtain infrared spectra. The most popular kind of spectrometer is called a Fourier-transform infrared spectrometer (FTIR).

As a chemical analysis technique, there are some advantages of FTIR such as:

1. FTIR can analyze different kinds of samples such as gasses, liquids, solids, semisolids, powders, and polymers.
2. FTIR presents rich information, the positions of peaks, their intensity, widths, and shapes in a spectrum which all supply useful information.
3. FTIR is an easy and fast technique. The majority of sample can be scanned in less than 5 min.

On the other hand, some disadvantages can be attributed to FTIR that are:

1. FTIR instruments are unable to measure aqueous solutions.
2. By using FTIR complex mixture cannot be analyzed.

In this research all sample measurement was done using PerkinElmer Spectrum 100 series with universal attenuated total reflectance (ATR) accessory. Actually, FTIR spectrometer work with controlled an interactive pressure so the ample contact with the diamond well and as a result the good quality of spectra can be obtained.

The ATR accessory works by calculating the variations that occur in a completely internally reflected infrared beam when the beam comes into contact with a sample. An infrared beam is focused onto an optically dense crystal. This internal reflectance makes an evanescent wave that expands beyond the surface of the crystal into the sample keeping in contact with the crystal. The evanescent wave protrudes just a few microns (0.5–5 μm) beyond the crystal surface and into the sample.

After that the sample absorbs energy in regions of the IR spectrum and the evanescent wave will be altered. Then, the attenuated energy of each evanescent wave will come back to the IR beam and then exit to the opposite end of the crystal and pass to the detector in the IR spectrometer.

After that, the system creates an IR spectrum. In fact bulk solid and powder are generally best analyzed on the single reflection ATR accessories, with diamond being a suggested and preferred choice for most applications because of its durability and robustness. After cleaning the crystal and collecting the background, the sample is located onto the small crystal area, followed by the pressure over the sample. In the spectrum 100, Universal ATR accessory, the pressure locks into a precise position above the diamond surface.

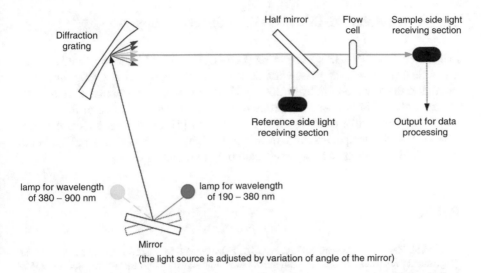

Fig. 3.5 Schematic diagram for principle of UV-visible spectroscopy and steps of spectra data collection

A "Preview Mode" is used in PerkinElmer Spectrum 100 series software that lets the quality of the spectrum to be monitored in real time while good tuning the exerted force.

3.6   Field Emission Scanning Electron Microscopy (FESEM)

The microstructure of all sample-based glass-ceramic was investigated using field emission scanning electron microscopy (FESEM). The FESEM analysis presents the sintered surface and fracture surface including grain size and shape. In FESEM electrons created by a field emission source are accelerated in a field gradient. The beam passes through electromagnetic lenses, focusing onto the specimen. As result of this bombardment, different kinds of electrons are emitted from the specimen. A detector catches the secondary electrons, and an image of the sample surface is constructed by comparing the intensity of these secondary electrons to the scanning primary electron beam. Finally, the image is showed on a monitor.

Microstructural observations were performed using a field emission scanning electron microscope FEI NOVA NanoSEM 230 machine equipped with energy-dispersive X-ray spectrometer (EDX), and the grain size was measured by the means of linear intercept method. To make the samples conductive for electron beam analysis, their surface was covered by a thin layer of gold at 50–70 Torr.

3.7   Ultraviolet-Visible (UV-VIS) Spectroscopy

In this study, the UV-VIS spectrometer (Lambda35, PerkinElmer) was used to study the optical properties of glass-ceramic samples. The absorption signal was measured for the wavelength from 200 to 1800 nm. It was presumed that for this research, the basis of absorption edge of the glass-ceramics is due to the indirect transition. The optical bandgap energy is given by [5]. Figure 3.5 depicts the principle of UV-visible spectroscopy in which the light from the lamp is radiated onto the diffraction grating and dispersed according to wavelength.

References

1. M.H.M. Zaid, K.A. Matori, S.H.A. Aziz, A. Zakaria, M.S.M. Ghazali, Effect of ZnO on the physical properties and optical band gap of soda lime silicate glass. Int. J. Mol. Sci. **13**, 7550–7558 (2012)
2. Y. Syono, S. Akimoto, Y. Matsui, Crystallization kinetic of glass particles prepared from a mixture of coal ash and soda-lime cullet glass. J. Solid State Chem. **3**, 369–380 (1971)
3. V.S. Aigbodion, J.O. Agunsoye, V. Kalu, F. Asuke, S. Ola, Microstructure and mechanical properties of ceramic composites. J. Miner. Mater. Charact. Eng. **9**, 528–538 (2010)
4. A.W. Coleman et al., Dehalogenation of binuclear arene-ruthenium complexes: a new route to homonuclear triruthenium and heteronuclear ruthenium-iron cluster complexes containing chelating phosphorus ligands. Crystal structure of Ru3(CO)10(Ph2PCH2PPh2). Inorg. Chem. **23**(7), 952–956 (1984)
5. R.A. Smith, *Semiconductors Second Edit* (Cambridge University Press, Cambridge, 1978)

Chapter 4
Result and Discussion

4.1 Introduction

This chapter presents the experimental results gained from the density and linear shrinkage measurement, XRD analysis, FTIR, FESEM analyses, and UV-VIS analysis. Thirty-six glass-ceramic samples were successfully prepared. All the samples were homogenous.

4.2 Density and Linear Shrinkage Analysis

To know about the molecular packing inside the glass-ceramic samples, the density of them was calculated. Figure 4.1 shows the density plotted as a function of sintering temperature for willemite doped with 5 wt.% Er_2O_3. Based on the information presented in the graph, it reveals an increase in density with increasing sintering temperature, with the highest density of 3.915 g/cm³ for the sample sintered at 1100 °C for 4 h, suggesting that rapid densification occurred at a temperature above 1000 °C. It is mainly due to the decrease in total fractional porosity of the sample with the increase in sintering temperature [1]. This value is in agreement with the theoretical density of 3.90 g/cm³. Hence the density achieved using the solid-state melting and quenching method was found to be 86.41% of the average theoretical density, and it is quite important that such a high level of density can be achieved without any sintering aid. This can be attributed to a large number of contact points formed because of the powder size and the higher fraction of fine particles resulting in shorter diffusion path. In such conditions, the average coordination number of particles will be increased leading to enhanced sintering [2]. Powder compressibility is

© The Author(s), under exclusive licence to Springer Nature Switzerland AG 2019
G. V. Sarrigani, I. S. Amiri, *Willemite-Based Glass Ceramic Doped by Different Percentage of Erbium Oxide and Sintered in Temperature of 500-1100C*, SpringerBriefs in Electrical and Computer Engineering, https://doi.org/10.1007/978-3-030-10644-7_4

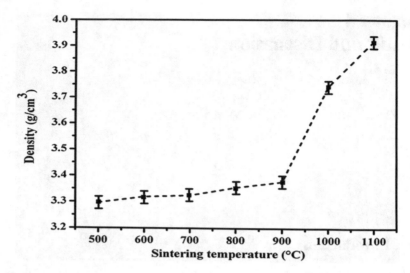

Fig. 4.1 The variation of bulk density and sintering temperature for willemite doped with 5 wt.% Er^{3+} powders

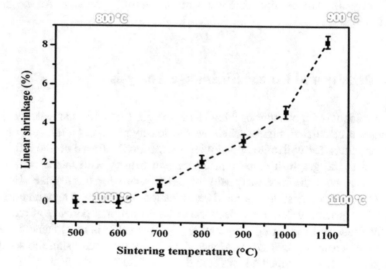

Fig. 4.2 The variation in linear shrinkage in the temperature of 500–1000 °C

considered as one of the most crucial factors in producing dense ceramics that directly affect the fired density of the final product. The compaction mechanism of brittle powders in rigid die is typically considered in three stages including sliding and rearrangement of the particles, fragmentation of brittle solids, and elastic deformation of bulk compacted powders [3]. Figures 4.2 and 4.3 show the result of linear shrinkage of pellets in relation to sintering temperature. There is significant increase of linear shrinkage with an increase in sintering temperature.

More result of density and linear shrinkage can be found in appendices section.

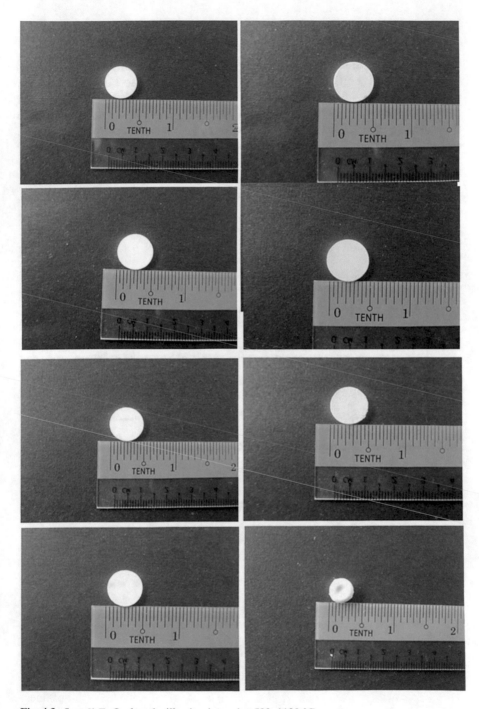

Fig. 4.3 5 wt.% Er₂O₃-doped willemite sintered at 500–1100 °C

4.3 XRD Analysis

The XRD patterns of Er_2O_3 (pure), glass frits, willemite (1000 °C), and Zn_2SiO_4:Er_2O_3 samples sintered at different temperatures ranging from 500 to 1100 °C for 4 h are presented in Figs. 4.4, 4.5, 4.6, 4.7, 4.8, 4.9, 4.10, 4.11, 4.12, and 4.13. The first pattern that depicts Er_2O_3 (pure) indicated that the main peak of pure erbium oxide is located at $2\theta = 29.29°$, and the second one illustrates the amorphous phase of glass frits. The third diffraction pattern shows willemite-based glass-ceramics. As it can be observed, it yielded six diffraction peaks overlapped on the diffuse peak of the crystalline Zn_2SiO_4. The diffraction peaks at $2\theta = 22.14°$, $25.61°$, $31.61°$, $34.11°$, $38.90°$, $49.01°$, and $65.87°$ corresponding planes which are (3 0 0), (2 2 0), (1 1 3), (4 1 0), (2 2 3), (3 3 3), and (2 2 6) can be indexed using standard diffraction pattern

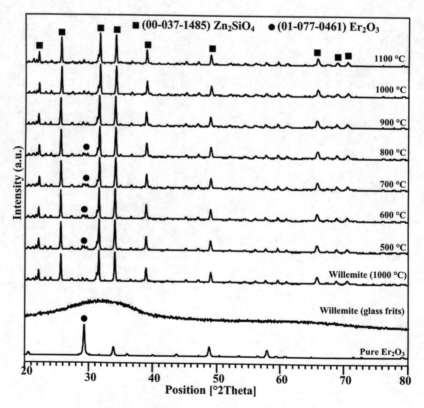

Fig. 4.4 1 wt.% Er_2O_3-doped willemite sintered at 500–1100 °C

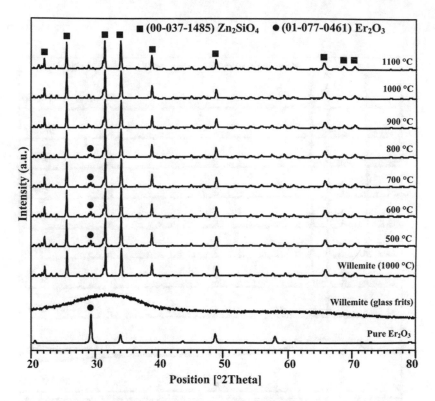

Fig. 4.5 2 wt.% Er_2O_3-doped willemite sintered at 500–1100 °C

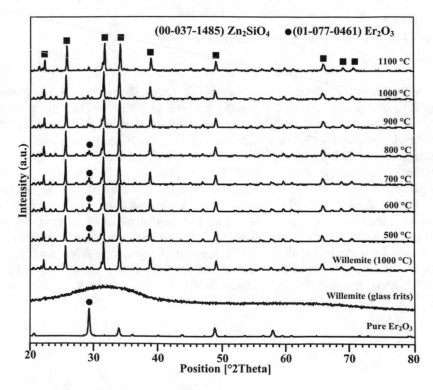

Fig. 4.6 3 wt.% Er_2O_3-doped willemite sintered at 500–1100 °C

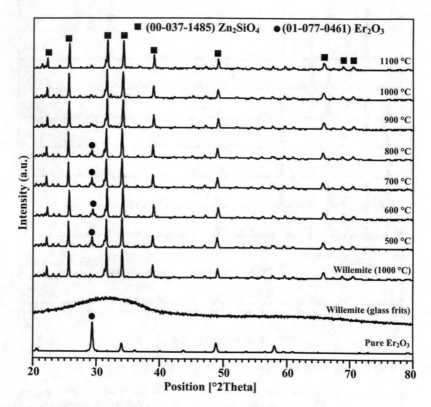

Fig. 4.7 4 wt.% Er_2O_3-doped willemite sintered at 500–1100 °C

of willemite crystal phase, respectively [4–6]. This indicates that rhombohedral crystalline willemite is formed by mixing ZnO and SLS glass and heat treatment at 1000 °C as an optimum temperature to produce willemite-based glass-ceramics [7]. When the prepared willemite was doped with 1–5 wt.% Er_2O_3 and sintered at a temperature of 500 °C, beside the diffraction peaks assigned to the Zn_2SiO_4, one additional peak is detected. The new diffraction peak at $2\theta = 29.29°$ can be indexed as plane (2 2 2) diffraction of cubic crystalline Er_2O_3. This is ascribed to the fact that the Er_2O_3 did not react with crystalline Zn_2SiO_4 and then produce a new phase in this sintering temperature. After that, looking at the patterns, it can be seen that the diffraction peak intensity at $2\theta = 29.29°$ tends to decrease gently from 500 to 700 °C followed by a dramatic fall in 800 °C sintering temperature. After sintering at 900 °C, the XRD peaks of Er_2O_3 diffraction peak completely disappears, and the XRD pattern has only the diffraction peaks of willemite without any additional

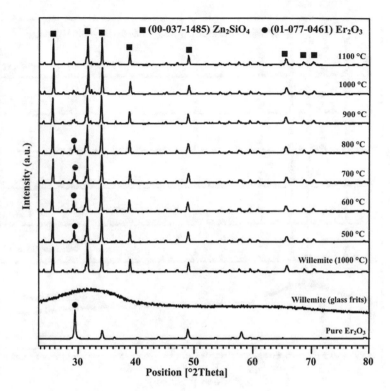

Fig. 4.8   5 wt.% Er_2O_3-doped willemite sintered at 500–1100 °C

diffraction peak seen. It can be concluded that the solid-state reaction between well-crystallized willemite and Er^{3+} can be obtained at 900 °C sintering temperature and Er^{3+} can completely dissolve in the lattice at this temperature. The phase formation of willemite-based glass-ceramic was further confirmed by FTIR. Kaewwiset et al. [8] reported that Er_2O_3-doped SLS glass was prepared by solid-state reaction at 1200 °C. Beside that [9] reported that manganese-doped zinc silicate green phosphor was produced with appropriate oxides by a solid-state reaction at a temperature of 1400 °C. The phase formation of willemite-based glass-ceramic was further confirmed by FTIR.

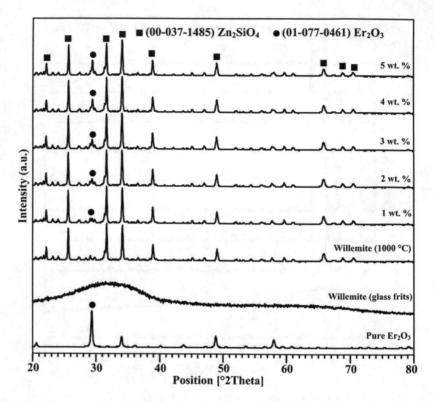

Fig. 4.9 Different contents of Er$_2$O$_3$-doped willemite sintered at of 500 °C

4.4 Structural Analysis by IR Spectra

FTIR spectroscopy was performed to obtain fundamental information concerning the functional groups of the studied glass-ceramic. The experimental FTIR spectra of the willemite-based glass-ceramics with the content of 1–5 wt.% Er$_2$O$_3$ at different sintering temperatures are presented in Figs. 4.14, 4.15, 4.16, 4.17, and 4.18. The bands detected in these FTIR spectra are summarized in Table 4.1. The assignments were made by comparing the experimental data gained in the present work with related vitreous and crystalline compounds. The FTIR spectrum of the glass-ceramic matrix, Zn$_2$SiO$_4$:Er$_2$O$_3$, consists of eight wide and strong transmission bands positioned at 462 cm^{-1}, 513 cm^{-1}, 576 cm^{-1}, 615 cm^{-1}, 702 cm^{-1}, 868 cm^{-1}, 904 cm^{-1}, 931 cm^{-1}, and 976 cm^{-1}. Normally, the metal oxide vibrations occur below 1000 cm^{-1} [10]. The transmission peak of lower wave number at 462 cm^{-1} is assigned to the totally symmetric stretching of SiO$_4$ group [11–15]. The band

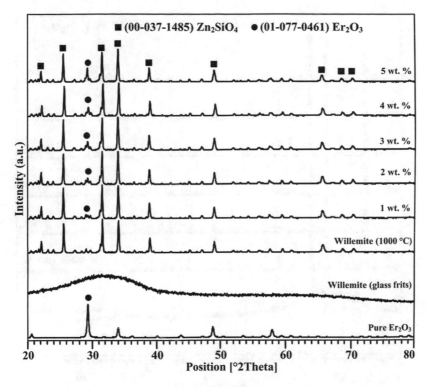

Fig. 4.10   Different contents of Er_2O_3-doped willemite sintered at 600 °C

Table 4.1 FTIR absorption features and their assignments for the $[(ZnO)\ 0.5\ (SLS)\ _{0.5}]_{1-x}\ [Er_2O_3]_x$

Wavenumber (cm^{-1})	Assignment of vibrational mode
4W-463	Si–O symmetric stretching vibration in SIO_4 units
513-518	Si–O bending vibrations
575-577	Zn–O symmetric stretching vibration in ZnO_4 units
614-616	Zn–O asymmetric stretching vibration in ZnO_4 units
697-699	Si–O bond vibration
868-869	Si–O symmetric stretching vibration
900-932-977	Si–O asymmetric stretching vibration

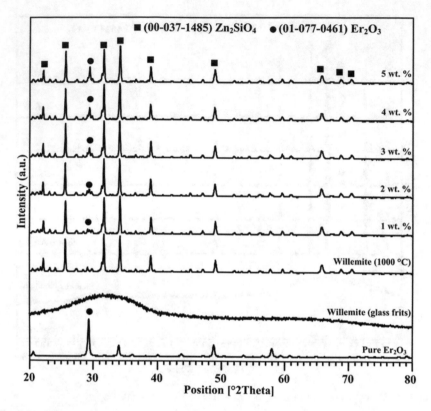

Fig. 4.11 Different contents of Er_2O_3-doped willemite sintered at of 700 °C

positioned at 513 cm⁻¹ corresponds to Si–O bending vibration modes [16, 17]. The peak located at 576 cm⁻¹ belongs to the symmetric stretching vibration of the ZnO_4 group [18]. The peak at 615 cm⁻¹ was marked as asymmetric stretching vibration ZnO_4 group [11, 13]. The band placed at 702 cm⁻¹ can be reported as Si–O bond vibration [19–23]. The transmission band situated at 868 cm⁻¹ corresponds to the symmetric stretching vibration of the Si–O [11, 13]. The broad bands located at 904 cm⁻¹, 931 cm⁻¹, and 980 cm⁻¹ correspond to asymmetric stretching of the Si–O [11, 13]. Increase in sintering temperature from 500 to 1100 °C increases the intensity of the IR bands. This is due to an increase of the structural order associated to the crystallization occurring in the heat-treated samples and agrees to the X-ray diffraction data [24]. It can be concluded that the compositional evaluation of the FTIR properties of the $[(ZnO)_{0.5}(SLS)_{0.5}]_{1-x}[Er_2O_3]_x$ system proposes that the presence of erbium ions affects the surrounding of the Si–O and trivalent erbium occupies its position and this is in agreement with the presence of Er^{3+} peak at 20.29 in

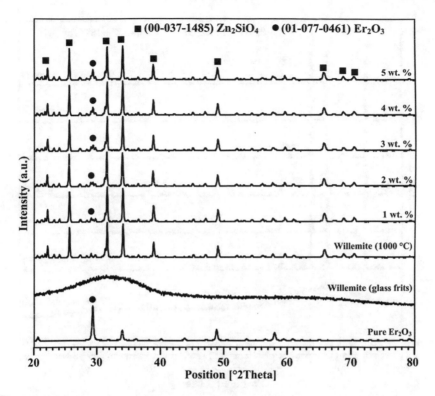

Fig. 4.12   Different contents of Er_2O_3-doped willemite sintered at 800 °C

the XRD pattern [24]. The FTIR spectrum of the crystalline (cubic) Er_2O_3 shows characteristic transmission bands located at 463 cm^{-1} allocated to Er–O bond vibrations [25]. Hence, the band at 462 cm^{-1} is due to the vibrations of the Er–O group present in the studied glass-ceramic system [24]. The most significant modification produced by the addition of erbium and increase of the heat treatment temperature of the studied samples is that it shows a drop in the intensity of FTIR band located at 513 cm^{-1}. This indicates that addition of erbium oxide and increase in sintering temperature reduced the presence of SiO_4 group.

4.5   Surface Morphological Analysis

Figures 4.19a–d, 4.20a–d, and 4.21 show the FESEM images for willemite, Er_2O_3, and 5 wt.% Er^{3+}-doped willemite. As it can be observed in Fig. 4.19a–d, the crystallized particles of willemite aggregated and were irregular, and additionally the

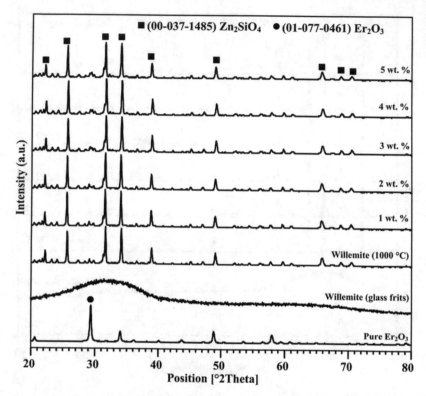

Fig. 4.13 Different contents of Er$_2$O$_3$-doped willemite sintered at 900 °C

willemite surface has a homogenous distribution with rhombohedral-like particles. Beside that Fig. 4.20a–d shows the particle of pure Er$_2$O$_3$. Looking at Fig. 4.21, it can be observed that when willemite was doped with 5 wt.% Er^{+3} and sintered at 70 °C, the Er^{3+} particles have dispersed on the surface of the willemite and have not occupied on the lattice. On the other hand, after increasing the sintering temperature up to 800 °C and 900 °C (Fig. 4.21), the surface morphology of the impact powder becomes granular, and it appeared to have a homogenous distribution. This indicated that Er^{3+} reacts with the willemite. However, image of Fig. 4.21 at 1000 °C shows that by sintering at 1000 °C, the agglomeration of willemite happened and the clustered Er^{3+} remains on the surface. Sintering at 1100 °C (Fig. 4.22 at 1100 °C), willemites completely cover the dopant. As a conclusion it can be seen that

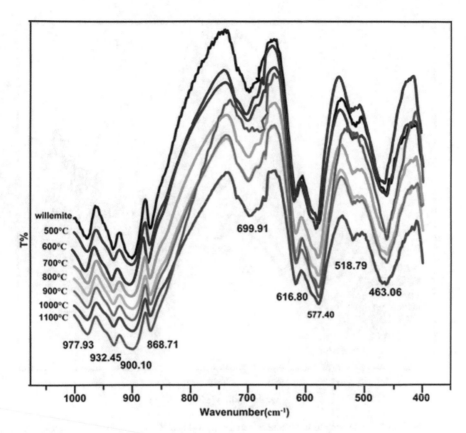

Fig. 4.14   1 wt.% Er_2O_3-doped willemite sintered at 500–1100 °C

amorphous phase reduced during sintering, and as the temperature increased to 1000 °C, less grain boundaries would be present due to grain growth, but at the temperature of 1100 °C, it is impossible to calculate the sample grain size due to melting of the sample. The distribution of samples was shown in Figs. 4.22, 4.23, and 4.24 and Table 4.2. The distribution of Er^{3+} in the willemite matrix was confirmed by FESEM/EDX analysis as shown in Figs. 4.22, 4.23, and 4.24. The EDX spectrum (Fig. 4.25) shows clearly the Zn, O, Si, Al, and Er peaks, in addition to the Au peak, resulting due to the sample being sputtered with gold for the conduction for clear viewing, while the Al peak is due to the SLS glass from waste material. The sintering temperature is on or above 1200 °C [9, 26] in most of the recently published investigations, which has been brought lower (900 °C) in the present work.

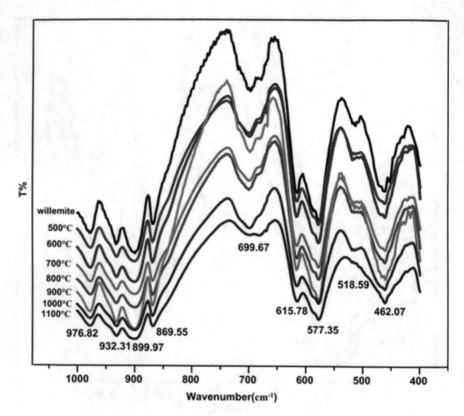

Fig. 4.15   2 wt.% Er$_2$O$_3$-doped willemite sintered at 500–1100 °C

Table 4.2   Table of particle
size of willemite doped with
5 wt.% Er$_2$O$_3$ sintered at
500–1100 °C

Sample	Grain size (nm)
Willemite	**604.71**
700	342.21
800	455.54
900	506.96
1000	625.2

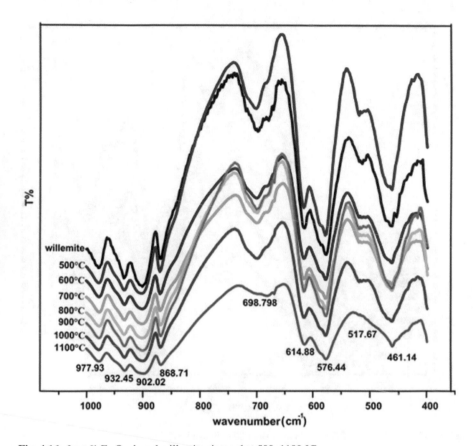

Fig. 4.16   3 wt.% Er$_2$O$_3$-doped willemite sintered at 500–1100 °C

4.6   Optical Property Analysis

In order to study the presence and optical performance of Er^{3+}-doped willemite and the effect of sintering temperature on the optical property of sample, absorption measurement was conducted, and the spectra are presented in Figs. 4.27 and 4.28. Figure 4.26 illustrates grand-state absorption transitions in erbium that are already confirmed by [19]. The UV-VIS spectrum of willemite doped with different percentages of Er^{3+} and sintered at the temperature of 500–1100 °C presented in Figs. 4.27 and 4.28 reveals the existence of broad and strong absorption lines located at about 1533, 980, 815, 650, 523, and 489 nm, corresponding to optical transitions

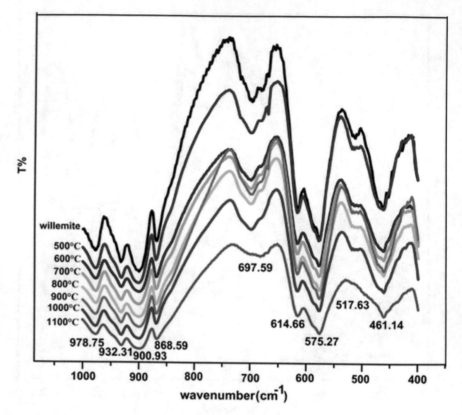

Fig. 4.17   4 wt.% Er_2O_3-doped willemite sintered at 500–1100 °C

from $^4I_{15/2}$ state to $^4I_{13/2}$, $^4I_{11/2}$, $^4I_{9/2}$, $^4F_{9/2}$, $^4S_{3/2}$, and $^2H_{11/2}$. The sharp absorption bands in the visible and the near-infrared domains correspond to the 4f–4f erbium electronic transitions. On the other hand, the band centered at 373 nm can be attributed to the host glass-ceramic.

The study of the UV-VIS spectra (Figs. 4.27 and 4.28) depicts that by increasing Er_2O_3 and sintering temperature, the characteristic UV-VIS bands are modified, namely:

1. All spectra of the doped glass-ceramics depict broad absorption band due to host matrix network with an intense feature situated at about 373 nm. The intensity of this band tends to grow by increasing the Er_2O_3 content in the range of 1–5 wt.%

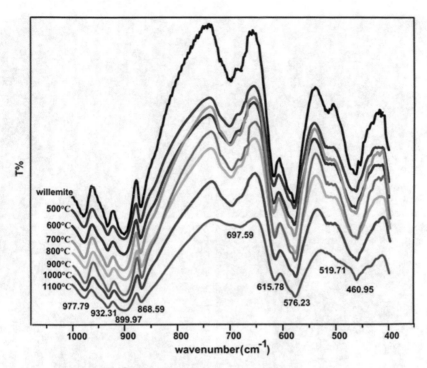

Fig. 4.18 5 wt.% Er_2O_3-doped willemite sintered at 500–1100 °C

and sintered at 500–900 °C range, followed by a drop in the temperature of 1000 and 1100 °C. According to ligand field theory [27], these glass-ceramics could illustrate charge-transfer band in the UV area due to the absorption ligands (oxygen atoms) around the cations, which generally are located in the range of 300–400 nm. Charge-transfer bands (O–Zn) could occur when a valence electron is transferred from the oxygen atom toward the occupied orbitals of the zinc cation, and those are parity allowed and yield generally in the UV and vacuum UV regions. The existence of the transfer charge between O and Zn is possible. The increase of the intensity of the IR band situated at about 560 cm^{-1} can be due to the vibrations of the Zn–O bond in the [Zn O_n] structural units.

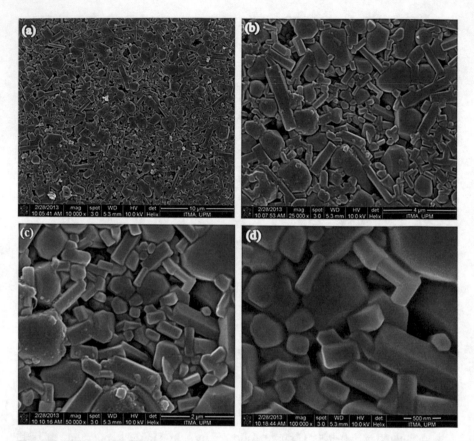

Fig. 4.19 Willemite in different magnifications, (**a**) 10 k, (**b**)-25 k, (**c**) 50 k, and (**d**) 100 k

2. By adding the Er_2O_3 content to the host network, and increasing the sintering temperature from 500 to 900 °C, the intensity of UV-VIS bands situated between 400 and 1800 nm tends to increase due to the absorption of Er^{3+} ions and the host crystal structure.

3. By increasing the sintering temperature to 1000 and 1100 °C, the intensity of UV bands tends to drop. It can be indicated that by increasing temperature at 1000 and 1100 °C, the Er_2O_3 particles tend to produce cluster that causes the decrease in the UV absorption bands.

4. For the sample with $x = 5$ wt.% Er_2O_3, two strong absorption bands situated at about 1535 and 523 nm were observed. These bands can be attributed to the optical transition from $^4I_{15/2}$ to $^4I_{13/2}$ and $^4S_{3/2}$ state, respectively.

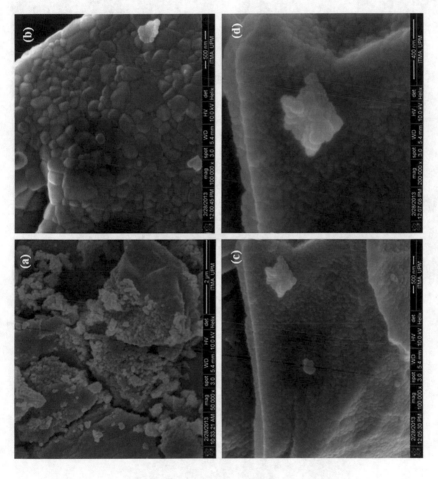

Fig. 4.20 Er_2O_3 in different magnifications, **(a)** 50 k, **(b)** 100 k, **(c)** 100 k, and **(d)** 200 k

Fig. 4.21 Pictures of 5 wt.% Er$_2$O$_3$-doped willemite sintered at 700–1100 °C

Model	Gauss	
Equation	y=y0 + (A/(w*sqrt(PI/2)))*exp(-2*((x-xc)/w)^2)	
Reduced Chi-Sqr	52.88892	
Adj. R-Square	0.62861	
		Value	Standard Error
Frequency	y0	25.45428	5.4512
	xc	662.57162	55.80511
	w	291.49598	149.48632
	A	-9400.56613	5419.95551
	Sigma	145.74799	
	FWHM	343.21029	
	Height	-25.73129	

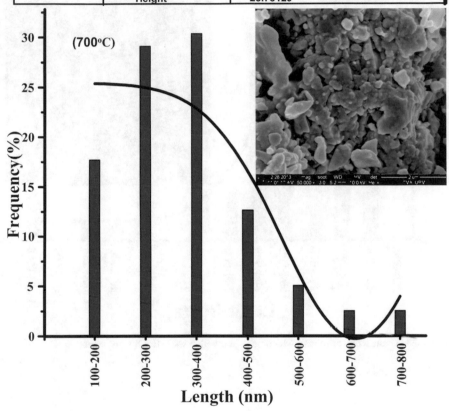

Fig. 4.22   Grain size distribution willemite doped with 5 wt.% Er_2O_3 sintered at 700 °C

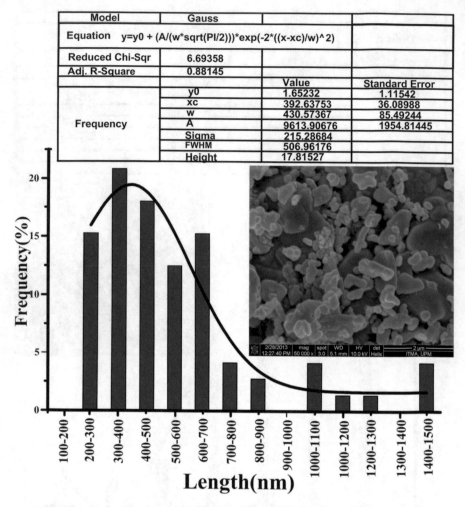

Model	Gauss		
Equation	$y=y0 + (A/(w*sqrt(PI/2)))*exp(-2*((x-xc)/w)^2)$		
Reduced Chi-Sqr	6.69358		
Adj. R-Square	0.88145		
		Value	Standard Error
	y0	1.65232	1.11542
	xc	392.63753	36.08988
	w	430.57367	85.49244
Frequency	A	9613.90676	1954.81445
	Sigma	215.28684	
	FWHM	506.96176	
	Height	17.81527	

Fig. 4.23 Grain size distribution willemite doped with 5 wt.% Er_2O_3 sintered at 900 °C

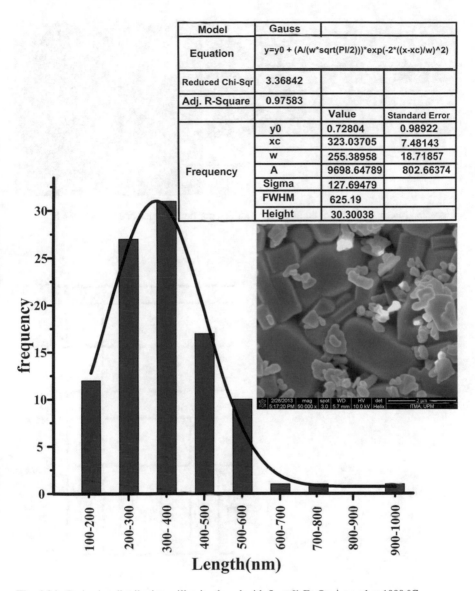

Model	Gauss	
Equation	y=y0 + (A/(w*sqrt(PI/2)))*exp(-2*((x-xc)/w)^2)	
Reduced Chi-Sqr	3.36842	
Adj. R-Square	0.97583	
		Value	Standard Error
	y0	0.72804	0.98922
	xc	323.03705	7.48143
	w	255.38958	18.71857
Frequency	A	9698.64789	802.66374
	Sigma	127.69479	
	FWHM	625.19	
	Height	30.30038	

Fig. 4.24   Grain size distribution willemite doped with 5 wt.% Er_2O_3 sintered at 1000 °C

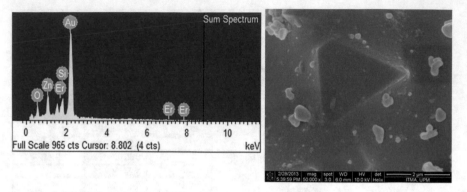

Fig. 4.25 Microstructural willemite doped with Er^{3+} powders sintered at 1100 °C for 4 h

Fig. 4.26 Grand-state
absorption transitions in
erbium

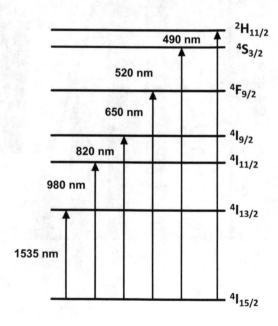

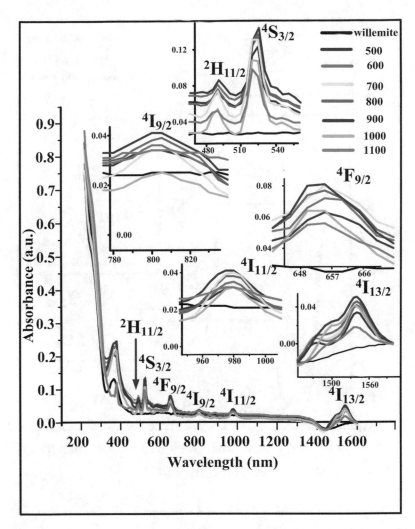

Fig. 4.27 Room temperature absorption spectra of 5 wt.% Er₂O₃-doped willemite sintered at 500–1100 °C

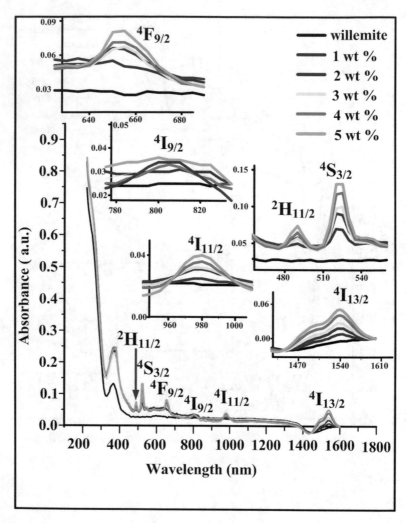

Fig. 4.28 Room temperature absorption spectra of 1–5 wt.% Er₂O₃-doped willemite sintered at 900 °C

References

1. Y. Guo, H. Ohsato, K. Kakimoto, Characterization and dielectric behavior of willemite and TiO₂-doped willemite ceramics at millimeter-wave frequency. J. Eur. Ceram. Soc. **26**(10–11), 1827–1830 (2006)
2. M. Mazaheri, A.M. Zahedi, M. Haghighatzadeh, S.K. Sadrnezhaad, Sintering of titania nanoceramic: densification and grain growth. Ceram. Int. **35**, 685–691 (2009)
3. P. Bowen, C. Carry, From powders to sintered pieces: forming, transformations and sintering of nanostructured ceramic oxides. Powder Technol. **128**(2–3), 248–255 (2002)
4. F. Marumo, Y. Syono, Effects of soda–lime–silica waste glass on transition of Er³⁺ formation kinetics and micro-structures development in vitreous ceramics. Acta Crystallogr. B **27**, 1868–1870 (1971)

5. Q. Lu, P. Wang, J. Li, Structure and luminescence properties of Mn-doped Zn_2SiO_4 prepared with extracted mesoporous silica. Mater. Res. Bull. **46**(6), 791–795 (2011)
6. H. Rooksby, A. McKeag, The effect of nucleation catalysts on crystallization of aluminosilicate. Trans. Faraday Soc. **37**, 308–311 (1941)
7. M. Takesue, H. Hayashi, R. Smith Jr., Thermal and chemical methods for producing zinc silicate (willemite): a review. Prog. Cryst. Growth Charact. Mater. **55**(3–4), 98–124 (2009)
8. W. Kaewwiset et al., ESR and spectral studies of Er^{3+} ions in soda-lime silicate glass. Phys. B. **409**, 24–29 (2013)
9. T. Cho, H. Chang, Preparation and characterizations of Zn_2SiO_4:Mn green phosphors. Ceram. Int. **29**(6), 611–618 (2003)
10. S. Kanagesan, S. Jesurani, R. Velmurugan, C. Kumar, T. Kalaivani, Magnetic hysteresis property of barium hexaferrite using D-fructose as a fuel. J. Mater. Sci. Eng. **4**, 88–92 (2010)
11. S.R. Lukić, D.M. Petrović, M.D. Dramićanin, M. Mitrić, L. Đačanin, Optical and structural properties of Zn_2SiO_4:Mn^{2+} green phosphor nanoparticles obtained by a polymer-assisted sol–gel method. Scr. Mater. **58**(8), 655–658 (2008)
12. Q. Chen, M. Ferraris, D. Milanese, Y. Menke, E. Monchiero, G. Perrone, Novel Er-doped PbO and B_2O_3 based glasses: investigation of quantum efficiency and non-radiative transition probability for 1.5 μm broadband emission fluorescence. J. Non-Cryst. Solids **324**(1–2), 12–20 (2003)
13. A. Shaim, M. Et-tabirou, L. Montagne, G. Palavit, Role of bismuth and titanium in Na_2O–Bi_2O_3–TiO_2–P_2O_5 glasses and a model of structural units. Mater. Res. Bull. **37**(15), 2459–2466 (2002)
14. R. Iordanova, Y. Dimitriev, V. Dimitrov, S. Kassabov, D. Klissurski, Glass formation and structure in the V_2O_5-Bi2O3-Fe2O3 glasses. J. Non-Cryst. Solids **204**(2), 141–150 (1996)
15. G.L.J. Trettenhahn, G.E. Nauer, A. Neckel, Vibrational spectroscopy on the PbO-$PbSO_4$ system and some related compounds: part 1. Fundamentals, infrared and Raman spectroscopy. Vib. Spectrosc. **5**(1), 85–100 (1993)
16. M.R. Ahsan, M.G. Mortuza, Infrared spectra of $xCaO(1-x-z)SiO_2zP_2O_5$ glasses. J. Non-Cryst. Solids **351**(27–29), 2333–2340 (2005)
17. D.R. Bosomworth, H. Hayes, A.R.L. Spray, G.D. Watkins, Proc. R. Soc. London Ser. B **317**(5), 133 (1970)
18. J. Lin, D.U. Sänger, M. Mennig, K. Bärner, Sol–gel deposition and characterization of Mn^{2+}-doped silicate phosphor films. Thin Solid Films **360**(1–2), 39–45 (2000)
19. A.J. Kenyon, Recent developments in rare-earth doped materials for optoelectronics. Prog. Quantum Electron. **26**(4–5), 225–284 (2002)
20. P. Capek, M. Mika, J. Oswald, P. Tresnakova, L. Salavcova, O. Kolek, J. Spirkova, Effect of divalent cations on properties of Er^{3+}-doped silicate glasses. Opt. Mater. **27**(2), 331–336 (2004)
21. E.F. Chillcce, I.O. Mazali, O.L. Alves, L.C. Barbosa, Optical and physical properties of Er^{3+}-doped oxy-fluoride tellurite glasses. Opt. Mater. **33**(3), 389–396 (2011)
22. A. Okamoto, T. Inasaki, I. Saito, Synthesis and ESR studies of nitronyl nitroxide-tethered oligodeoxynucleotides. Tetrahedron Lett. **46**(5), 791–795 (2005)
23. G. Will, W. Pies, A. Weiss, Simple Oxo-compounds of Silicon Without H_2O, NH_3,(Simple Silicates), in *Key Element: Si. Part 1*, ed. by K. H. Hellwege, A. M. Hellwege, vol. 7d1a, (Springer, New York, 1985), pp. 60–79
24. M. Bosca, L. Pop, G. Borodi, P. Pascuta, E. Culea, XRD and FTIR structural investigations of erbium-doped bismuth-lead-silver glasses and glass ceramics. J. Alloys Compd. **479**(1–2), 579–582 (2009)
25. F.F. Bentley, L.D. Smithson, A. L. Rozek, *Infrared Spectra and Characteristic Frequencies 700–300 cm^{-1}*, vol. 68, no. 1 (Interscience Pub., New York, 1968), pp. 103–188.
26. V. Sivakumar, A. Lakshmanan, S. Kalpana, R. Sangeetha Rani, R. Satheesh Kumar, M.T. Jose, Low-temperature synthesis of Zn_2SiO_4:Mn green photoluminescence phosphor. J. Lumin. **132**(8), 1917–1920 (2012)
27. B.N. Figgis, *Introduction to Ligand Field Theory* (Wiley, New York, 1966)

Chapter 5
Conclusion and Future Research in the Glass-Ceramic Field

5.1 Conclusion

The effect of different percentages of Er_2O_3 (1–5 wt.%) and different sintering temperatures ranging from 500 to 1100 °C on the structure of the samples was investigated using x-ray diffraction (XRD). The XRD results indicate that rhombohedral crystalline willemite is formed by mixing of ZnO and soda lime silicate (SLS) glass and heat treatment of the mixture at 1000 °C. They also show that when willemite (Zn_2SiO_4) is doped with different percentages of erbium oxide (Er_2O_3), one additional peak belonging to Er^{3+} appears at a sintering temperature of 900 °C; the additional peak disappears because of the reaction between willemite and Er^{3+}. The Fourier transform infrared spectroscopy (FTIR) bands show that the appearance of the vibrations of the SiO_4 and ZnO_4 groups at 462 cm^{-1}, 513 cm^{-1}, 576 cm^{-1}, 615 cm^{-1}, 702 cm^{-1}, 868 cm^{-1}, 904 cm^{-1}, 931 cm^{-1}, and 976 cm^{-1} clearly suggests the formation of a Zn_2SiO_4 phase. The compositional evaluation of the FTIR properties of the $[(ZnO)_{0.5}(SLS)_{0.5}]_{1-x}[Er_2O_3]_x$ system proposes that the presence of erbium ions affects the surroundings of the Si–O bond and trivalent erbium occupies their position. This is in agreement with the appearance of the XRD peak positioned at 29.29°. The FTIR spectrum of the crystalline (cubic) Er_2O_3 shows characteristic transmission bands located at 462 cm^{-1}, assigned to the Er–O bond vibrations. The density calculation indicates that the highest density is 3.915 g/cm^3 in the sample sintered at 1100 °C for 4 h, suggesting that rapid densification occurs at a temperature above 1000 °C, firstly because of the decrease in the total fractional porosity of the sample with the increase in sintering temperature, and secondly because of formation of a large number of contact points due to the microstructure of the powder. The linear shrinkage results show that with an increase in the sintering temperature, the linear shrinkage of the samples tends to increase, indicating that a higher sintering temperature results in more compacted glass-ceramic particles and tends to

© The Author(s), under exclusive licence to Springer Nature Switzerland AG 2019

G. V. Sarrigani, I. S. Amiri, *Willemite-Based Glass Ceramic Doped by Different Percentage of Erbium Oxide and Sintered in Temperature of 500-1100C*, SpringerBriefs in Electrical and Computer Engineering, https://doi.org/10.1007/978-3-030-10644-7_5

decrease the porosity while increasing the density, as well as the linear shrinkage, of the glass ceramic. The surface morphology of the samples was analyzed using field emission scanning electron microscopy (FE-SEM) followed by investigation of particles in the samples by use of energy dispersive x-ray spectrometry (EDX). The obtained data show that when willemite is doped with different percentages of Er^{3+} and sintered at 500–700 °C, the Er^{3+} particles disperse on the surface of the willemite and do not occupy the lattice. On the other hand, after an increase in the sintering temperature to 800 °C and 900 °C, the surface morphology of the impact powder becomes granular and it appears to have a homogeneous distribution. This indicates that the Er^{3+} reacts with the willemite. However, with sintering at 1000 °C, agglomeration of willemite happens and the clustered Er^{3+} remains on the surface. With sintering at 1100 °C, the willemite completely covers the dopant. The optical properties of the samples were studied using ultraviolet–visible spectroscopy (UV-VIS) spectra. The study of the UV-VIS spectra showed that the UV-VIS spectrum of willemite doped with different percentages of Er^{3+} and sintered at a temperature of 500–1100 °C reveals the existence of broad and strong absorption lines located at about 1533, 980, 815, 650, 523, and 489 nm, corresponding to optical transitions from the $^4I_{15/2}$ state to $^4I_{13/2}$, $^4I_{11/2}$, $^4I_{9/2}$, $^4F_{9/2}$, $^4S_{3/2}$, and $^2H_{11/2}$. The sharp absorption bands in the visible and near-infrared domains correspond to the 4f–4f erbium electronic transitions. A band centered at 373 nm can be attributed to the host glass ceramic. These results indicate that the sintering temperature has a significant influence on the shape of the crystals in the investigated glass-ceramic samples. In fact, on the basis of these results, it can be concluded that the optimum sintering temperature for willemite doped with Er_2O_3 is 900 °C.

5.2   Future Work

The current research has successfully elucidated the structural, physical, and optical properties of willemite glass ceramic doped with Er_2O_3. This work can be extended in two different ways:

1. Besides Er_2O_3, another kind of rare earth (such as Eu_2O_3 or Y_2O_3) or a metal element can be used as a co-dopant. The physical, structural, and optical properties of such a combination can then be investigated.
2. The photoluminescence properties of the current and proposed samples can be studied, and with use of the photoluminescence data and the Judd–Ofelt theory, the transition probabilities or oscillator strengths for the various transitions between energy levels in the rare-earth ions can be calculated.

Appendix

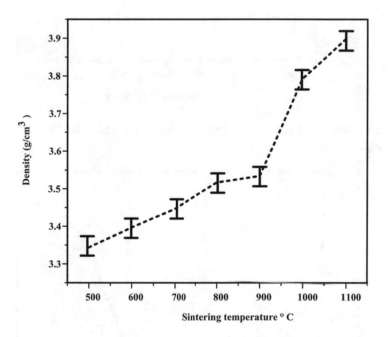

The variation of bulk density and sintering temperature for Willemite doped 1 wt.% Er^{3+} powders.

© The Author(s), under exclusive licence to Springer Nature Switzerland AG 2019
G. V. Sarrigani, I. S. Amiri, *Willemite-Based Glass Ceramic Doped by Different Percentage of Erbium Oxide and Sintered in Temperature of 500-1100C*, SpringerBriefs in Electrical and Computer Engineering, https://doi.org/10.1007/978-3-030-10644-7

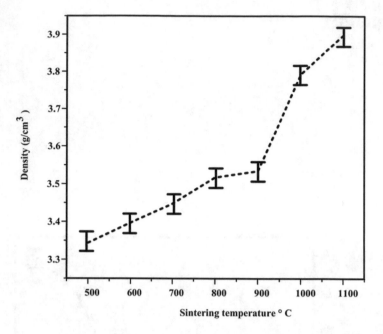

The variation of bulk density and sintering temperature for Willemite doped 1 wt.% Er^{3+} powders

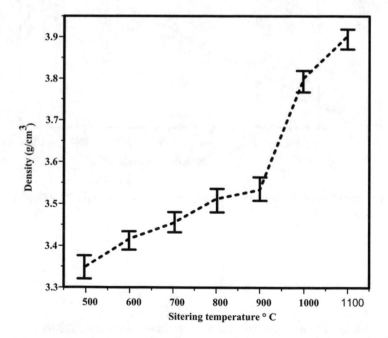

The variation of bulk density and sintering temperature for Willemite doped 3 wt.% Er^{3+} powders

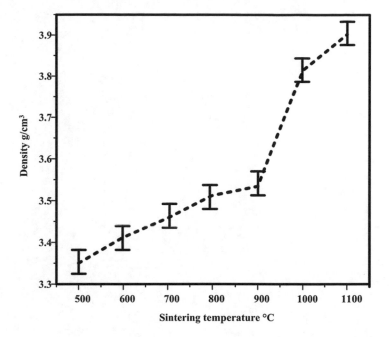

The variation of bulk density and sintering temperature for Willemite doped 4 wt.% Er^{3+} powders

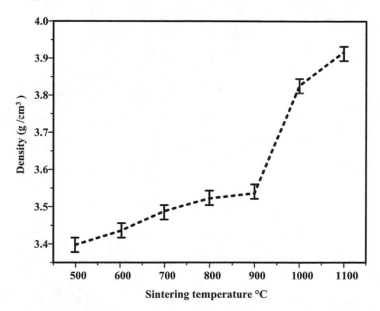

The variation of bulk density and sintering temperature for Willemite doped 5 wt.% Er^{3+} powders

Index

© The Author(s), under exclusive licence to Springer Nature Switzerland AG 2019   71
G. V. Sarrigani, I. S. Amiri, *Willemite-Based Glass Ceramic Doped by Different
Percentage of Erbium Oxide and Sintered in Temperature of 500-1100C*,
SpringerBriefs in Electrical and Computer Engineering,
https://doi.org/10.1007/978-3-030-10644-7

Printed in the United States
By Bookmasters